ARMS CA$H

Lost in the Hall of Mirrors

Memoirs of a Defence Salesman

RED ROCK
BOOK WRITERS

by
Ralph Houston

'Every gun that is made, every warship launched, every rocket fired signifies in the final sense, a theft from those who hunger and are not fed, those who are cold and are not clothed. This world in arms is not spending money alone. It is spending the sweat of its laborer's, the genius of its scientists and the hopes of its children. This is not a way of life at all in any true sense. Under the clouds of war, it is humanity hanging on a cross of iron.'

from an address by **Dwight D Eisenhower**
April 16, 1953

Chapters

Chapter 1
Into the Hall of Mirrors

The call came in on my hotel room 'phone at around 05:30 hrs and disturbed my brief but much needed sleep. I wanted to ignore the ringing, but it was too persistent to be ignored, so I reached out reluctantly and grabbed the handset. I was physically exhausted from too many flights in quick succession to and from the Middle East, and I was mentally drained also following the rancorous argument which exploded and raged in my Gulf hotel Suite a few hours earlier. The physical debris from this jarring early morning meeting remained in-situ; partially emptied pots of now lukewarm coffee, dirty cups and glasses, half-eaten Room-Service snacks and ashtrays piled high with cigarette butts, mostly mine, which spilled over onto the glass table-top. The air was rank and stale with the reek of Marlboro smoke and the meeting with my potential 'Agent' had been rank and stale too. It ended badly.

I'd first met this aspiring Agent, a former Arab 'Ambassador Extraordinary and Plenipotentiary' some months earlier, since when we'd engaged in the standard Defence Company courtship ritual by meeting secretly in hotels around the Gulf to discuss a potential Defence deal in his home country. At each meeting we considered how we might co-operate and collaborate on an Arms deal in which the former Ambassador proposed himself as my Company's 'Agent'. In this capacity his reward would be Commission, a kickback, call it what you will: Arms Cash. Such payments would have to be strictly covert, so the need for personal and corporate security was high. My employers, the National Defence Company (NDC), badly wanted and needed the lucrative sales revenue and the former Ambassador certainly wanted and desperately needed the Arms Cash. Neither party viewed this proposition as being in any way crooked, this was simply how business was conducted in the Defence domain and it played out to a well-established code, an accepted modus operandi which was held securely in check by an Industry Omerta.

This clandestine commercial flirting added thousands of miles to my personal Air Miles account and my Expenses tab soared well above

1

its already high level, but this mattered not a jot to NDC, since on the Corporate expenditure scale my budget was simply petty-cash in NDC's great scheme of things. My running costs were paltry compared to the potential Arms deal, worth around fifty million pounds initially with more lucrative stages to follow, which I was gently nudging onto the table with my Middle Eastern Customer.

As NDC's Middle East and Pakistan Marketing Executive I was on familiar ground here both geographically and professionally; I managed these territories because I had hard-earned form in Defence Markets with a contract-winning track record. I'd previously served in both the British and Oman Land Forces and spoke some Arabic too, but such military experience, although helpful sometimes, was not the critical ingredient in Defence Sales success. Instead, it was the combination deep local knowledge, commercial and Sales savvy, and the willingness to accept risks which were the truly valuable tools in the box.

It doesn't follow that ex-military personnel are automatically suitable for international Defence Sales roles; that is a common misconception, and in my own experience the reverse more often applies. Without hard-edged sales and commercial skills, ex-military personnel can become liabilities very quickly, and the more senior they are, the more pompous and inflexible they are likely to be. These folks are generally too rigid and opinionated to be successful in international Sales environments, where business is realised not in the simple black and white terms with which they are comfortable, but rather, in a thousand shades of grey which vary frequently according to myriad factors and moods. The ex-military bull-in-a-china-shop model doesn't work so well because Customer and Agency hot-buttons change frequently, and great agility is needed. The only constants are Competitors, Secrets and Arms Cash.

The characteristics of the successful international Defence Salesman are undetectable by any number of fashionable HR psychometric tests, and the good ones, and there are very few of these, are seriously unbalanced people. How else could anyone do these jobs, why would anyone do these jobs? In Middle Eastern markets every Sales gambit under the sun, every Carpetbagger's trick and every ploy has been tried and tested and all are familiar to the regional Arms Buyers and Agents who matter. Whatever you think you bring to the

Arms Bazaar Party it's unlikely to surprise anyone. The very best of them bring 'the something else' into the arena, not just product knowledge and feckless optimism, but guile and a creative empathetic ability to see and to create possibilities. You need canniness to structure these deals and the willingness and nerve-lessness to haggle on a grand scale and to bluff when necessary. My job was to sell hard-core Defence products in an unforgiving and highly competitive marketplace and I was very good at it.

The process of setting-up and attending covert meetings with potential Arms Cash Agents, such as the acrimonious early morning affair with the former Ambassador, was bread and butter to me. It was a key component in my job with a major western Defence Plc which targeted Agents before Sales, on the basis that if you got the right Agent the Sale was guaranteed; well, that was the prevailing theory. But that early morning call I took in my drowsiness differed to most others I might receive for it brought with it potentially fatal consequences.

"You're in mortal danger…" my anonymous caller said "…get to your Embassy immediately…someone in the City wants you dead." This caller seemed to know all about the row which had raged so aggressively and so bitterly just a few hours earlier and knew too about the potential Defence deal with its lucrative but illicit Arms Cash rewards. It seemed a bit surreal and in my slow waking-state I prompted my Caller to repeat his message. He was insistent, he was adamant, I was in mortal danger, someone was going to kill me. The ex-Ambassador was the obvious candidate, but in the Hall of Mirrors you never really know who is working with whom, it could be anyone.

Travelling alone and selling or dealing in any way in lethal Defence products you must take warnings like this very seriously, especially where large Arms Cash commissions are at stake, for international Defence Sales have a bloody history. In 1993 the body of the Head of Procurement for the Taiwanese Navy, Captain Ying Chin-fen, was found floating off the coast of Taiwan, triggering an investigation into bribes in an Arms deal involving the French Company Thomson-CSF, in what became known internationally as the *Lafayette Affair*. A Coroner determined that the Captain had been murdered and it was later reported that irregularities he'd been investigating related to the purchase of Frigates, which had resulted from the bribing of Naval Officials. Arms Cash. The vessels themselves were below the standard

3

anticipated by the Users, but product quality hadn't been the key component of the sale, Arms Cash was. An investigation resulted in major French Defence Company Thales, which succeeded Thomson-CSF in name, being ordered by a Swiss tribunal to pay c. $850m in damages and interest and in several Naval Officers being jailed. It was thought that the deceased Officer had been about to expose the Arms Cash in the deal, but this intention cost him his life and other suspicious deaths followed.

When huge sums of Arms Cash are at stake anything goes. A key witness and participant in the Nigerian 'Armsgate' scandal was assassinated, and Investigators into the corrupt deal for Agosta Class submarines in Pakistan in the '90's, involving some $4.3m worth of Arms Cash, speculated that the non-payment of the full amount of kickbacks may have led to the murder of French nationals in the 2002 terrorist attack in Karachi, although that was largely hearsay. The Daily Telegraph reported in 2017 that former French Prime Minister Edouard Balladur had been charged in connection with the Arms Cash arrangements in the deal.

The very process of selling Defence materiel can be as lethal as the products or systems themselves. In 2004, freelance US Arms Dealer Dale Stoffel was shot dead in Iraq after he alerted US Government personnel to Defence-related kickbacks in the Baghdad Green Zone. Stoffel was no novice, but despite his self-cultivated soldier-of-fortune image, he might have known better than to raise his profile whilst still in the field, especially in Baghdad at that time, as no one is going to help you out in such circumstances. Defence Sales Freelancers and Corporate Executives circulate in similar orbits of course, but different rules apply. Both are disposable of course, but Freelancers more so. Dale, RIP brother.

Defence Sales can be a very solitary occupation, unlike Military duty when there is usually a co-protective team of others around. If things do go awry, you can bet your bottom-dollar you'll be on your own, and my personal experiences in working for major and minor international Defence companies confirms this. Big Plc's especially, are quick to engage reverse gear when trouble looms. Being shot at or bombed as a serving soldier is the luck of the draw, call it Karma, Allah Kareem, Destiny, whatever you want, the Shooter or the Bomber doesn't know you personally and doesn't care who you are, you're just

4

a target. It was your misfortune, your fate maybe or your own stupidity perhaps, to be standing in the crosshairs of the Sight or on some buried IED when it was detonated. Bad orders, bad timing or bad tactical positioning on your part maybe, but it's not personal. Your enemies simply want to kill; all of you preferably, but any one of you will do. Be consoled, you haven't been singled out for special treatment. My early morning threat warning was different though, this was personal, I was singled out.

International Defence business is inherently risky, even without threats from participants, as the statistical probability of unplanned occurrences is high and is multiplied by the irregularity and secrecy of the work. Frequent international travel invites disruption and flight hazards, whilst health risks, particularly in remote or hostile regions, are standard. Misunderstandings with covert Agents and overt Customers over money can get heated very quickly, and even potential Coups can interfere with progress. No User-Manual is available to cover all this, you just have to improvise. I was alone in the Middle East with a lucrative Defence deal on the table, and someone I didn't know had warned me that I was targeted for assassination, specifically, for 'chopping-up' en route to the Airport as the Caller so indelicately put it.

I needed to respond quickly but thoughtfully, what exactly should I do now, what could I do? My options were complicated by my schedule, the pressing element in which was a formal meeting with senior Officers and Procurement Officials at 0800hrs that very morning in the HQ Procurement Office. This meeting was now just a couple of hours away, and maybe in different circumstances I could have postponed it with an excuse and re-arranged my Programme. But this special appointment was the critical next stage in the pursuit of the major deal I was positioning and to cancel it at such short notice would create significant difficulties, it could jeopardize everything.

It had taken months of maneuvering and patience to assemble this morning's Meeting group and re-arranging everything would be a lengthy, if not impossible task, in the short-term, made more difficult by the onset of Ramadan which was imminent. All Military Offices would be closed for business during the Ramadan period so I would lose the momentum, the tempo of the deal, and Competitors could and would close-in. Equally, I was hardly eager to dwell in the City to present a soft target to some anonymous assassin.

Discounting competitive pressures and debilitating intra-Company politics, Defence deals are invariably complex and difficult to close for many reasons, most of which arise outside the Customer domain. In those days the foremost amongst these was the identification, selection, validation and management of Arms Cash Agents, who could facilitate contracts by providing privileged information or influence, and who would enable covert access to key Insiders. The reward for these services was the payment of Arms Cash commissions by the successful Company, through suitably indirect and untraceable channels; increasingly awkward in the Digital age, though not impossible to achieve as new circumstances have bred new solutions.

The early morning threat added an unwelcome layer of complication to the existing mountain of issues which had to be surmounted in pursuit of a signed contract. The warning landed with all the subtlety of an exploding hand grenade. What was going on? I'd pressed the Caller for more information as he claimed he was passing on this warning, not making a threat himself. So, who was making the threat? I pushed him for more, but he sounded quite nervous, as if he'd gone out on a limb to alert me. It was like a hostage negotiation and I needed to keep him talking but he just wouldn't or couldn't elaborate. He was adamant though, I was in mortal danger, and he told me repeatedly that I was to be snatched and chopped-up. He urged me to get to the Embassy without delay. He hung-up.

Was this a scare tactic, a bit of theatre perhaps, or a credible threat? I'd have to decide. I knew that working secretly in the Middle East with high financial stakes would raise the ante, but was it enough to kill for? The compellingly earnest tone in the Caller's voice suggested that yes, someone thought it was, and in truth I knew it to be so. I visualized being stopped on the Freeway, perhaps being dragged out of a taxi, stabbed maybe, or shot, possibly beheaded. I imagined different scenarios and mentally rehearsed how I might react or resist. I recalled examples of reported abductions to remind myself of the importance of reacting as quickly as possible, since once dragged down an incident pipeline, escape becomes firstly more difficult, and then to all intents and purposes impossible.

No one would come to the rescue so my response to any incident would have to be determined, aggressive and quick, there would be no second chance. I was an easy target and there was enough at stake in the deal I was positioning, enough Arms Cash, to make someone greedy enough to kill for it. Financially it wasn't on the same grand scale as say the infamous Airbus Bribery case, but it was enough. The Airbus case was the subject of this US Department of Justice Press Release dated 31st Jan 2020:

'Airbus SE (Airbus or the Company), a global provider of civilian and military aircraft based in France, has agreed to pay combined penalties of more than $3.9 billion to resolve foreign bribery charges with authorities in the United States, France and the United Kingdom arising out of the Company's scheme to use third-party business partners to bribe government officials, as well as non-governmental airline executives, around the world and to resolve the Company's violation of the Arms Export Control Act (AECA) and its implementing regulations, the International Traffic in Arms Regulations (ITAR), in the United States. This is the largest global foreign bribery resolution to date.'

I wasn't operating on this grand scale, but I opted to believe the Caller anyway; it would have been madness to do otherwise. Sometime after the event I learned that my Caller was a distant member of the old Afghan Royal family, information which would have been meaningless to me then as I knew nothing about that dynasty. Mohammed Zahir Shah was the last King, the last 'Shah' of Afghanistan, and he had abdicated after a forty-year reign following a coup in 1973. Afghanistan was obviously a 'no-go' zone for me even before the idiotic Western intervention and humiliating exit. The Mujahedeen had seen-off the Russians, the Taliban saw-off the West and their ancestors had seen-off Queen Victoria's Army; these folks were well practised. So, it seemed especially ironic that an Afghani had called, from Singapore as I discovered later, to save the life of a Western Defence Salesman in the Middle East. Odd things happen in this Industry, but this was right up there with the oddest.

I had a simple choice to make. I was just an employee and this wasn't a military duty; was I in or out? Soldiers swear Oaths to serve their King and Country or whatever, in exchange for peanuts and frequently for abandonment by Society if subsequently wounded, disabled, or suffering from PTSD. Military duty includes and embraces lethal risk as a central part of the job; it's something you accept, something you actually seek-out when you sign on the dotted line and open yourself to chance or to destiny in some foreign place. But putting one's life at risk over an Arms Cash treasure chest for a Plc? That was different. How much risk was I willing to take?

Ex-military Company employees usually carry with them their militaristic senses of duty and loyalty, but these are truly burdens, for these characteristics are highly valued and much exploited by Employers, especially in the Defence Sector. Ex-military personnel will willingly go the extra mile or carry the extra load, they are victims. It's in their nature to do so and they can't help it. Scorpions and Frogs. I did that too, I was always 'Mission focussed' regardless of the circumstances, but maybe this time I should reset, after all, I was the one at risk.

On later trips into the Region, and mindful of abductions and murders in Iraq, I reflected on just how close I might have come to a premature and messy death right there and then, and all in the interests of Company Business. Shareholders never faced these problems when cashing-in their Dividends. Was taking a gamble reasonable or worthwhile I asked myself, the Company certainly didn't give a toss or even pay very well. NDC's Corporate Directors were interested primarily in counting-out their bonuses and we all knew that actual leadership was not their forte. How far was I prepared to push my luck? The Seventh Cavalry wasn't going to gallop magically over the hill, bugles blaring, to my rescue.

The clock was running down to my critically important HQ appointment, so now it was a simple 'go/no go' question. In that moment I made a quick military-style 'Combat Appreciation' to distil my options. Tactics Courses had taught me this simple mental process which helps Commanders work out what to do next. I made my choice. I was a long way from home and without support, but the bigger the risk the sharper the senses, the keener the appetite, so I decided to push the envelope. Frankly speaking it was a 'Rush'. Run a f***ing Corporate Psychometric test on that.

I decided to attend the scheduled meeting despite the threat, but I'd take whatever evasive or deceptive actions I could to mitigate the risks. I'd also try and enlist some local help. I couldn't afford either for the deal or my sensitive approach-work to it to blow-up in my face whilst I was still in the Customer Country, that would be in nobody's interests, least of all mine. A Media revelation that a British-based Defence Company Executive was clandestinely engaging with local

Defence Procurement Officials or senior Military Officers for Arms Cash co-operation would be dynamite locally and internationally. At home the Transparency and anti-Arms Lobbies would have a field day. What happened subsequently would depend upon how much risk I was willing to take, as neither my NDC employers nor the Magic Fairy were going to help out, and the Embassy was ill-equipped and probably unwilling, as it later proved to be, to deal with such occurrences. It was up to me, which was probably best.

I 'phoned Reception to advise that I'd be checking-out later and requested my bill, then I temporarily slipped out of the Hotel via the fire-escape stairs into the adjoining Shopping Mall which formed part of the hotel complex. Despite the early hour, I was able to buy a cheap plastic suitcase from an Asian Vendor and returned with it to my room, where I shovelled 'room stuff' towels and such-like into it to give it weight, and I tagged the case with an airline baggage tag (always carry a few spares, ideally for different airlines).

I usually travelled with just a carry-on sized bag to speed my progress through Customs and Immigration channels and to avoid tedious passenger queues. Surely there should be a special slick channel reserved for Defence Company Sales Executives. I was a trained exponent in the art of the 'Business-Class Dash', that grumpy Executive exit from a passenger jet, and the subsequent sprint to and through Customs, which enables ever-testy businessmen to stay ahead of the travelling Masses. Only on the longest of long-haul trips did I have to succumb to the unwelcome encumbrance and nightmare of Hold-baggage. Oh! the indignity of waiting at luggage carousels with the great travelling unwashed and even tourists in some locations. The shame of it! My new plastic suitcase was part of an instant if somewhat flaky deception plan.

I paid my Room bill with my pre-authorised Corporate Amex Card and descended to the Lobby, where I made a royal performance of depositing my freshly-stuffed and falsely-tagged new suitcase with the Concierge for collection later in the day, which was a standard practise for late-departing travellers. I loudly announced the collection time for the benefit of any Listeners or Watchers, but of course I didn't plan to come back for it. I kept one favourite military adage in my head '…having limited options does not mean do nothing…' Free of the suitcase I headed back through the hotel Coffee Shop to the Shopping Mall, glancing around constantly.

The Mall was by then teeming with morning workers, hotel shift-changers and shop-openers. I didn't want to take a Taxi to the HQ from the exposed and highly observable front of the hotel, so instead I hailed a battered cab off the street. The HQ Office was about a twenty-minute drive from the Mall. My cab was a tatty saloon car and not a smart State-authorized Taxi, but I didn't notice the make, it was a Japanese something or other. The driver's Prayer beads dangled from the rear-view mirror and swayed hypnotically with the motion of the vehicle. Perhaps I should have been praying. Instead I hunched down rather comically in the back of the car whilst simultaneously trying to keep a view of the road via the Cab's wing mirrors, neither of which appeared to be aligned or adjusted in any useful way. They must have been as useless to the driver as they proved to be for me, although he wasn't bothered by this misalignment as he didn't use them.

The journey-time allowed me to review my situation yet again, although I'd done little else since the call came in. It was all weird. The originator of the threat had to be the Ambassador, who'd been incandescent with rage at the end of our meeting when he stormed out of my room. Where the Afghan caller fitted in was a mystery I couldn't solve. My decision not to work with the Ambassador would, in his eyes at least, have denied him a mouth-watering five to seven-million-dollar pay-day as NDC's Agent, and with two similar phases to be concluded over the next two or three years, each of a similar value, the ex-Ambassador could have had pencilled in a revenue stream of fifteen to twenty million dollars. It's nearly always the case that wannabe Arms Cash Agents who cannot really 'do the business' have mentally spent their Arms Cash before any contracts had even been drafted. By way of contrast, hardened Arms Cash Players are often serial participants who don't really need the money but are addicted to accruing more.

The whole process of getting a major Defence contract drafted, approved and signed is invariably a more complex and time-consuming process that most first-time Agents and home-based Directors think it's going to be; especially amongst Directors of Commercial Departments, the 'Never-Never-land' of the Defence Industry. There's inevitably Corporate frustration over the slow pace of contracting, but I don't know why anyone expects the commitment of a multi-million-dollar Defence budget to be quick or glitch-free. In this case it wasn't frustration over the slow pace of the deal which fuelled the former Ambassador's anger, but rather my decision not to appoint him as NDC's Agent, despite his extensive, expansive and expensive efforts to convince me to do so. Was his fury enough to make him want me dead? Probably it was. I could understand his angst, he'd seen his potential Arms Cash mountain tumble and crumble to dust before his very eyes and all because of me. But in the white heat of our early morning argument when I'd declined to engage him, I did so with very good reason.

Convincing though his personal credentials might be, the former Ambassador had not actually demonstrated his ability to secure the proposed deal, or more importantly to NDC at the time, to arrange the 'off-stage' engagement the Company wanted with the Head of Defence Procurement, who was a senior local Royal family member. Despite his claims of influence, arranging this highly desired and secret meeting seemed beyond the former Ambassador's reach.

The Ambassador did have something of value to offer though, as he'd been a source of detailed hard-copy data about the products and systems to be purchased. Even though this was 'Informant-level' material not 'Influencer' stuff in Arms Cash terms, it had a value to NDC. He was very coy about the true source of it. The Ambassador had postured as an Influencer, but in reality, he was somewhere between an Influencer and an Informant, peddling valuable but second-hand material. So, who was the 'First-hand' and how did I connect with him?

In the Hall of Mirrors there are gaping Arms Cash differences between Informers and Influencers. For modest-sized deals, engaging an Informer was often enough, but big deals needed real Influencers and they're much more difficult to engage with, indeed, the very process of engagement brings unthinkable Corporate risks of rejection and exposure. The high-quality documentary information stream, coupled with the Ambassador's undeniable persistence (oh! that UK Defence Salesmen had a fraction of his persistence) would undoubtedly have prompted NDC to accommodate him for something in the deal, a modest one-off payment perhaps, just as the Company might do for any other reliable Informant. But the ex-Ambassador would have to stop posing as an Influencer demanding an Influencer's reward and would have to reveal the true source of his data, trusting NDC, i.e. me, not to cut him out of the Arms Cash loop.

But as I'd discovered, the former Ambassador had been uniquely duplicitous even by the low-bar standards of the Defence Industry, and his actions placed him firmly outside the Arms Cash circle of trust.

Since I'd first become aware of this potential Requirement, I'd followed up over a dozen possible leads in a quest for the detailed information, data and influence I needed to get a winning Offer on the table and close a deal. I did of course continue to visit the Customer through the front door in the conventional manner to deliver product presentations and Proposals, and I conducted equipment demonstrations until I was blue in the face. All were welcomed by the Customer and politely received, but it would be stunningly naïve to believe that any Defence Company could or would actually win this business without a significant Agent and an Arms Cash deal in place. Arms Cash activities weren't unique to NDC as all major competitors did the same, it's just how things worked.

Formal Defence 'Requirements' exist at all points on the value scale from high-end Strategic Systems, which cost billions and take years to bring to fruition, to regular purchases of low-end consumables. Requirements may be stimulated by technological evolution, urgent operational needs, the availability of something new in the Marketplace, or just regular maintenance, and Defence Buyers procure Systems, Products and Services. Not all Defence deals are multi-billion-dollar events and not all National Defence Procurement processes are of equal sophistication. Requirements can be, and sometimes were of course fabricated solely as Arms Cash vehicles, just as Lord Stokes observed in his Sixties Report to the UK Government.

The best option from a Defence Company perspective, was to have a formal Customer Requirement shaped and written around its own Products, Systems or Services, and covertly supported within the Customer base to create a unique and unassailable winning proposition. Getting that done required serious Insider-help and co-operation which came at a price, an Arms Cash price. The merest whiff of a forthcoming Requirement triggered competing Arms Cash Agents to get to work on establishing their 'own' Companies whilst dislodging competitors. This activity ran far ahead of any formal Tendering announcements or Competitions (Competitions? lol!). The best antidote to all Arms Cash manoeuvres is to design, develop and market unique, 'must have' Defence products or systems, for which nothing need be paid to covert Agents, because the Customer will buy them regardless of anything. Just count-up how many F.16s are in service around for example.

Quarter is neither offered nor granted during the clandestine pre-positioning process, so individuals and organizations should anticipate serious commercial damage during the manoeuvring phase. A Company can expend a lot of Corporate resource chasing down a wet dream. Companies not in control of things are best advised to quit the field, as they are simply wasting their own time and Company money. In today's parlance these are 'No Bid' situations, as even before the Requirement becomes common knowledge, you already have no chance of winning. When some Defence System or equipment with no apparent suitability whatsoever for the buying Country's Forces is suddenly purchased, it simply indicates a satisfactory conclusion to someone's Arms Cash arrangements. You missed the boat.

Officials or Insiders who proactively seek Arms Cash deals with Defence Companies often team with local businessmen as their front men to become their external points of contact, their 'cut-outs' with a friendly Defence Company. Local Agents will also proactively cultivate Insider relationships, simply to create Arms Cash potential. Some Insiders who'd never considered such propositions may be persuaded to bend a little by local Agents with the promise of tsunamis of potential Arms Cash. It's often the Agent who approaches the Company in the first instance, and for National level deals a Ruling family member will probably be involved in many countries; certainly, Politicians will.

In my own experience, a surprisingly wide range of individuals in customer Countries make discreet approaches to international Defence Suppliers to alert them to opportunities and offer to smooth the passage of Contracts in exchange for Arms Cash. Defence Company Executives must be deft in sorting the Agent wheat from the chaff, as I had to with the former Ambassador, but this isn't always easy and expensive mistakes are made. The acid test as far as I was concerned was connectivity. Any self-proclaimed front-man claiming to operate on behalf of a Minister or senior Official, should be able to arrange a face-to-face meeting in a neutral location with his Principal when a large deal was at stake. If an Agent was unwilling or unable to arrange such a meeting, just walk away. For lower-level, lower-value deals it's much simpler of course, and the contact levels are easier to identify and engage.

A good Agent provides a firewall for his Principal and a safe communication channel for the Company, whilst a bad one gets too ambitious and tries to cut a deal directly with the Company. When this happens there were usually kickback offers for Company Executives to bind them in. A good Agent may be the working point of contact for a Defence Company, but all Agreements with major Influencers should only ever be concluded and signed with the Principal in person, at whatever level that happens to be, relative to the deal in hand. Many wannabe Agents demand the world on a plate, but it's important to stand firm and proceed only on a deal-by-deal basis to start with to build up confidence. Your Agent might be a dud.

In formal Bidding Competitions a broad spectrum of technical, financial and competitor information is required to land a contract, but only a fraction of the necessary data can be gained via official meetings with Procurement Staff on Customer sites. You really need to know that your Company is going to win before you bid. All Competitors in these situations know that a combination of information and influence is a pre-requisite for success, and as with the former Ambassador, as with many others, NDC was ready, willing and able to pay out big chunks of Arms Cash for what it really wanted. In this case they really wanted a tame conduit to the (Royal) Head of Defence Procurement. Therefore, 'closing' the current opportunity wasn't NDC's only objective, the Board wanted a whole lot more. The targeted ruling family individual enjoyed deep political influence, way above his token Defence Ministry appointment and day job; he was destined for the very top and the right relationship with him now would be of strategic and enduring value to NDC. The Cash Registers would ring and ring.

In NDC terms this could lead to major successive Technology, Product and Systems contracts across the Group for years ahead. Full employment played well with the UK Government, which never looked too closely at major deals, whilst massive bonuses and high Dividends played even better with Group Directors and Shareholders. NDC used people like me to create the possibilities and to get the Company 'in'. I kicked-down the doors and was focussed on near-term business. Once locked-on, the Corporation would take over and run the territory from the Centre and I'd be despatched to light another fire elsewhere. I knew my activities could unlock a golden future, both for NDC and the prime Influencer(s) in this State, and that's why I'd spent as much time as I had on the former Ambassador, he just might be the route in.

Equally, I got to kiss innumerable frogs during such quests, and I could almost feel the eyes of the Board swivelling towards me; would I find them the 'in' they so desperately sought? Left to their own devices they would struggle to find their way to the corner shop. Years earlier when I was less streetwise, I might have been hypnotised by the Ambassador's smooth patter and the appetising documents he wafted under my nose; I know many others were thus beguiled. In my early international days, I might less questioningly and more optimistically have offered him an Arms Cash deal straight off the bat, without probing more deeply into his capabilities or alleged connections. This had been a common habit in the Industry, but times had changed.

For all his excellent documents and valuable information, the ex-Ambassador was a few years too late with this type of 'see-it-and-buy-it' approach, as a good story and wads of documents were no longer enough. Even the Ambassador's personal offer, the Arms Cash Agent's standard kickback proposition to Executives, was not enough to persuade me to go with him. I knew the business potential for NDC extended way beyond this current requirement, large though it was, and I needed to add heavyweight strategic 'wasta' to NDC's Agency black books and in this Middle Eastern territory and only a strong Royal Family connection would cut it. Beyond the current deal lay a tantalizingly lucrative long-term future for NDC if I got this right, and the Corporation had much bigger, sexier and more expensive things to sell. NDC wanted a decisive high-level Connection, and death threat or no death threat, I was going to get them one. See how this military mission-focussed training f***s you up?

Connections, connections, connections, the Defence Industry, like most others, is all about connections. Nothing odd or unusual about that. The obvious difference is that Defence Industry Connections deal in a heady cocktail of State Secrets, lethal products and high-technology Systems, all lubricated with the smoothing, soothing flows of Arms Cash for which the former Ambassador thirsted. This potential deal had been no different from so many others to begin with, and although not on the uber-scale of say, Military aircraft, it was always going to be big relative to my own NDC Division's Systems.

But this Prospect had set Competitors a-flutter in an unusually agitated way and the chatter in the bazaars was intense and excited; there was something peculiar going on here but what was it? It wasn't the potential value, it was substantial yes, but there were bigger deals around for Companies which could afford to take them on, and it wasn't the location either, as this wealthy Middle Eastern Customer State was credible and well-established with plenty of Defence cash to spend. No, there was something else about it, but what was it?

I was ever mindful of the Iran-Iraq War Bull-market in illicit Arms shipments and wondered if any of this Requirement's materiel could be destined to pass through to an alternative destination. Infamous Lists of weapons and equipment 'Needs' had circulated discreetly and not so discreetly during the Iraq-Iran War, when a key issue had been one of false End-User certificates, so I trod very carefully. I'd once been questioned myself as to what I might have known about the trans-shipment of electronic Fuses from one seemingly genuine Middle Eastern buying State named on the End-User certificate, to Iran, but luckily it was all news to me!

In more recent times, German Sniper scopes which were sold to Iran via Italy, surfaced amongst Taliban militants fighting NATO troops in Afghanistan and nine people were charged with violating international embargoes. Today, the fashionable 'gifting' by the UK and USA of weapons and ammunition on a huge scale to Ukraine has messed-up the old-style End-User Protocols and will come back and bite us on day in the foreseeable future. Weapons 'donors' should always anticipate being shot at one day at with their own systems, the precedents are abundant. If it's become ok for the West to 'gift' weapons purchased by their own taxpayers for their own Forces to another State, then any other Country can and will purchase and divert weapons too when it suits them, and this will not always suit the West's playbook.

I'd exploited my own connections to cross-reference the ex-Ambassador's proposition, and my discoveries led to the acrimonious and potentially lethal bust-up in my hotel room. During that early morning meeting I finally confronted the Ambassador with copies of Agreements he'd made with leading international Defence Companies, including NDC's (i.e. my) sharpest competitors. I obtained these documents from a range of sources and contacts, some of whom sought NDC Agency appointments for themselves in their home

Countries. They presented me with the information as demonstrations of their personal credibility as potential NDC Arms Cash Agents. To say that it would be imprudent to ask how they obtained this material would be a f*** off understatement.

What you see is not necessarily what you get in the Hall of Mirrors and contrary to the Defence Industry's embedded obsession with discretion and secrecy, the ex-Ambassador had raised and sustained a very-high personal profile. Far too high in fact. He pedalled his story about how he could fix this Middle Eastern Defence business to Companies in Europe, USA, South America and Russia. He'd become hugely and badly over-exposed, and such exposure made Defence Executives like me very nervous. Arms Cash Rule Number One: Never attract unwanted attention. Arms Cash Rule Number 2: Never attract any attention. Like moths attracted fatally to flames, international Defence Sales and Marketing Executives were drawn-in by the prospect of the business apparently on offer via the former Ambassador. Major Companies took him at face value and signed-up with him as their Arms Cash Agent, unthinkingly, unquestioningly and with that feckless optimism which surmounts cold logical thought in so many Board Rooms. I'd convinced myself that he had a foot in more than one End-user camp too, but whose?

It took a long time to piece it all together. The Ambassador's story was very intricate and his activities criss-crossed Continents in which he'd contacted a multiplicity of Defence Companies. I exchanged confidences with my opposite numbers in competing Corporations, meeting-up in true Defence Industry style in random bars, restaurants, hotels and various other rendezvous points, even beside swimming pools and in a Hotel Gym in Dubai (Danni Minogue was touring at the time). I chatted with Agents and Arab businessmen I knew, and traded crumbs of information and fragments of Industry gossip, until finally it all became clear. Of course, the truism 'there's no such thing as a free-lunch' applied, so there had to be a little give and take in this intensive information collection and collation process. Little deals were done. As US General Tommy Franks once said in another context 'You have to trade...'

What finally emerged through the mist was that the Ambassador was betting on every horse in the race. He'd sought Arms Cash Agreements with every major Competitor and deployed the same convincing story and data to get our attention. He flirted with all of us and we all willingly flirted back, it was in our nature, we simply couldn't help ourselves. Scorpions and Frogs. I admired his gall as there was an almost breathtaking scale and ambition to his operation. Secrecy within the Industry ensured that he nearly got away with it, but the Defence business is a hostile and unforgiving domain in which to play such duplicitous games; tease the Hyenas, as the Ambassador had chosen to do, and sooner or later you'll get badly bitten; formidable jaws.

The documentation and data the Ambassador provided to his potential Arms Cash-paying patrons was unusually, almost uniquely seductive; it was exciting material, Defence Industry Porn. 'Treasure' in the Intelligence world. It consisted primarily of Inventory printouts listing in intimate detail the entire Defence arsenal of the host-country and showing Supplier, price and specification data with consumables re-order levels. The Ambassador fed us all selected morsels of this data and we snapped it up and hungered for more, it was like a feeding-frenzy at the Zoo. But data alone, however good, is not enough to land an Arms Cash Agreement or win a Contract, so this is where his Ambassadorial ambience, his 'Act', came into play. He came across so credibly and knowledgeably that one felt he could walk into any Ruler's Office in the Middle East and be greeted as a valued and trusted friend, and who knows, maybe he could?

The ex-Ambassador's Holy Grail was to engineer a series of Arms Cash paydays and he knew full-well that a Defence Company like NDC would be happy to pay Commissions for the Business he offered to facilitate, that was normal enough. But to make doubly certain of success, not only had the Ambassador almost signed-up the entire field, except NDC of course, but he'd also sent out seemingly unconnected 'Runners' carrying similar information, with each of them trying to secure secondary Agency Agreements. I loved the enterprise of it all, so creative.

His concept was akin to operating an international Arms Cash franchise, and this duplicated activity was not as contradictory as it might appear. Many, probably most of the Companies he approached, didn't rely on a single in-country Agent, Source or Feed for information or influence, but rather operated multiple Agents at different levels and in different domains, each supposedly insulated or unseen from the others. One Agent might manage, say, low-level spares or consumables like ammunition, with others handling project-specific or higher-level areas such as Command and Control Systems, Air Defence etc. Company obsessions with secrecy created conditions in which Players like the Ambassador could succeed, and trust me, Defence Companies were desperate enough for business and stupid enough to fall for this gambit.

Not all Agents were illicit, as some Countries allowed open Company representation akin to Distributors, and many large Defence Companies operated extensive regional Offices and technical Facilities with local Partners to support contracts overseas. It was also not the case that Companies always proactively sought out Arms Cash Agents, and sometimes Client's staff would discretely indicate their preference for a particular local Company or businessman as their chosen 'interface'. In most cases though a covert high-level Influencer would be essential for all strategic purchases, and that's what NDC wanted here. The ex-Ambassador did not fit that bill, but he had a role to play if he could open the right door and NDC would reward him for that.

Key Agent's activities were typically coordinated from a Corporate central point ('Marketing' lol) which sanitised and communicated inputs to operating Divisions to preserve the Agent's anonymity, so the number of Company personnel entrusted with Agency data was necessarily highly restricted. The higher the status of the Agent, the fewer Company personnel knew. To make life extra complicated, some Divisions within large Groups also ran tactical Agents of their own, when the Division felt it couldn't trust the 'official' Agent imposed by the Corporate Centre, or simply because no business was flowing via the sanctioned channel.

My experience of several Defence 'Primes' is that Divisional staff often had little faith in their corporate marketing teams, regarding them as remote and unnecessary overheads and frequently believing them, rightly or wrongly, to be on-the-take for kickbacks. The Ambassador had clearly identified these Corporate fissures and exploited them beautifully, and I could only applaud him for his comprehensive understanding of the 'System'. Unfortunately for me, I was all that stood between him and a Royal Flush of lucrative Arms Cash appointments with key Suppliers.

I'd been approached by nearly a dozen different 'Runners' each claiming to be able to fix a deal for NDC and each making claims about the strength of their contacts; their golden entry ticket to the Arms

Cash Club. The variety of approaches was bizarre and but for the lethal nature of the products, verged on comical. I was approached by an American Ranch Owner who boasted a share in a major Paris Nightclub, and we met for afternoon Tea in the Dorchester. He'd flown in from the US especially for this meeting.

We nibbled delicate sandwiches and sipped aromatic Tea like elderly tourists, before he reached into his briefcase and extracted, triumphantly, a photocopied and edited version of the Customer War Stock Programme, which unbeknown to him, I already possessed. I stayed silent as he briefed me in subdued tones amidst the chinking the afternoon Tea Service, on his ability to lead me to and through this deal; it was becoming a familiar script. He was crestfallen, when after listening carefully to him but saying little, I produced an identical document to his. He visibly glowed with anger at the revelation, having believed, as did the Ambassador's many other 'Runners,' that he was in possession of unique privileged and exclusive information for which NDC would pay a fortune. Wrong, oh! so wrong. The list of angry people disillusioned with the ex-Ambassador was growing.

To help my American friend recover from the aftershock of being duped, we left our Tea and headed-off for a real drink. I needed to establish what he knew, if anything, of any of the Ambassador's other Runners. I'd met an Indian businessman, a Swiss-based Lebanese and two Executives from other Defence Companies, which could not compete with NDC, but whose Executives wanted to be 'in' on the lucrative Arms Cash element of the deal. These latter two Jokers had sold their souls to the ex-Ambassador for kickbacks; everyone wanted a slice of this action.

I'd been approached directly by the Lebanese man, having been pre-warned about him by yet another Player in this increasingly complicated game. I was advised that he had 'off-centre' connections in Africa and he was described as being virtually a terrorist. This was probably dramatic; Players in this Industry are prone to be dramatic. But there was no way to corroborate or to confirm any of these suggestions or rumours. I met him in London initially, but uniquely amongst all the Runners, he also turned up one day at NDC's Midlands Divisional Offices demanding a meeting with me; an unheard-of action in the Agency world as no one ever comes to Company Offices. He was tenacious, I grant him that, and highly aggressive too, but I wasn't going to be bounced into giving him an Arms Cash Agency Agreement and he left in an unhappy and embittered state. Another one.

I was also contacted by two Competitor's Agents who wanted to switch sides, and who offered me personally something 'extra' in the deal if I would switch the NDC Agency to them. It was commonplace for hungry potential Agents to offer Regional Company Executives an Arms Cash back-hander, and Executives who took the bait bound themselves to their illicit Patrons, locked forever in the mutual essentiality of secrecy. Rumour has it that at least two major Defence Company CEOs had dipped sticky fingers into the Arms Cash Till and had to be discretely slipped out of the Industry. Some of the wannabe-Agents I met seemed credible enough at first glance, but none of them had the connectivity which would lead to the top, which is what NDC wanted.

The former Ambassador certainly put on a good show whenever we met, the ambience was carefully cultivated, and he had a great sense of theatre. He arrived for meetings in fashionable Gulf Hotels in his Arab Diplomatic robes looking for all-the-world like a serving member of a Government or like some Middle Eastern Royal figure, and this went down particularly well with Hotel Staffs, who were naturally deferential to him. His every wish and whim were catered for and there were whims a-plenty it seemed.

He was a genuinely seasoned international traveller of course, with useful contacts in the Diplomatic world. His script was smooth and well developed, it was honed. He'd acquired Defence subject-matter knowledge and spoke confidently about Systems, weapons and Defence materiel as if he was familiar with them in his everyday life. He gave the convincing impression that he was empowered to conclude a contract personally, although of course he wasn't. I could see how folks might be taken in by him so 'chapeau' for his Act.

The ex-Ambassador had been a Smoker earlier in his life but had quit, so now instead of smoking he would take a cigarette and roll it endlessly between his fingers during a conversation or meeting, sniffing the tobacco occasionally and putting the cigarette between his lips, sometimes even talking with it in place but never actually lighting it. During the course of a long meeting, and there were oh so many of these, the packed tobacco would loosen, finally falling in shreds from the cigarette paper. He would discard the empty paper and take another cigarette and repeat the ritual. He went through a huge number of my cigarettes in this way and covered many documents and coffee tables with unsmoked tobacco strands during our meetings.

After what seemed like an endless series of meetings with him and others, and from the chats I was having in the Bazaars, I knew that he could neither influence the Procurement decision I sought nor arrange the off-stage meetings I really wanted. He had a link into the Procurement Office, but it was not a senior connection, probably it was a tame Officer, a friend, a Clerk or similar, someone with access to the paperwork and programmes and this was common. Bold and entertaining though the Ambassador's show had been, I would have to tell him that it was over, his game was up, no deal no 'Arms Cash'.

The cigarette-sniffing former Ambassador aimed to ensure success by covering the Competitor field, but without NDC in his hand his suite of Companies was incomplete, and his plan would be fatally flawed. If the only Company with which he'd failed to secure an Agreement won the business he would get nothing, and his hugely expensive global excursions would have been in vain. And oh boy! had he travelled and spent some money. I had sight of his Amex bill once and the overdue debt was huge. Doubtless he had a long queue of other creditors, as he'd been making undeliverable promises in a dangerous domain and desperately needed an Agency Agreement with NDC.

In my hotel room in the early hours of that morning I finally challenged him about these Competitor Agreements and even showed him signed copies I'd obtained of two of the most conflicting ones. These documents bore his signature and yet he denied that they were his. It was a conspiracy against him he said, it was a trick, it was a pack of lies. Then suddenly he changed tack. It had been his cunning plan for both of us to get rich, and then he tacked again, he'd been compelled to do it by some powerful people with vested interests whose identities couldn't be revealed.

On and on it dragged. It was all gibberish of course and the argument escalated to a shouted vitriolic climax and his stormy exit from my suite. I'd plotted against him he said, I would pay a heavy price for that. I thought this was all bluster and frustration until I took the early morning 'phone death threat warning call. It was not the first time I'd had to manage difficult potential Agents but none of the others had, so far at least, threatened to kill me, although that was destined to change. Quite where the Singapore-based Afghan Caller featured in this odyssey remained a mystery. I peered into the Arms Cash Hall of Mirrors seeking answers, but only transient phantom faces reflected silently back at me.

It's in every Agent's interests to look after his Defence Company Handler and vice-versa, but there's no textbook solution for managing these relationships. Agents and Company Executives come with hugely differing temperaments and egos, and arguably the fact that they both operate in the Defence Industry indicates an absence of ethical distractions. Of the two broad categories of Agent, Informants and Influencers, Informants were the cheapest to reward and were used in medium to low-value deals, wherein simple pricing or specification data might be critical, whilst Influencers operated at much higher-value and more sophisticated levels. Identifying and engaging Informants was quite easy, Influencers much less so. Sometimes Informants were self-identifying!

Customer Support Staff in many Middle Eastern Defence Offices were expatriate Pakistanis, the Clerical Class of the Middle East, but also Somalis, Sudanese or Lebanese, the latter being regarded as the Business Managers of the region. I might be visiting a Headquarters or Technical Department perhaps, making repeat visits over time and I'd become a familiar figure. I tried to sustain an approachable and friendly posture with all Staff and Officers, in which respect being British is a positive advantage, since the ability to exchange trivial, meaningless conversation for hours on end is a national trait. Brits are also very good at just waiting for things to happen; another valuable characteristic in the Region. By way of contrast, Americans always want to make things happen in a hurry, they love deadlines, timetables and numerical measures on all things; sometimes this is appropriate and works, but often it doesn't.

Staff in Procurement Offices were fully aware of the value of the documents and data they handled and some fed information back to friendly local Businessmen, who in turn used it to tout their services as Arms Cash Agents. Maybe this was the case with the ex-Ambassador. Friendly Clerks in Customer's Offices would sometimes deliberately leave documents on view as indicators of their availability and willingness to engage and sometimes they were comically self-proclaiming in offering their services outside the Office environment.

Some would-be Informants even risked approaching Company Executives in their Hotels, either telephoning ahead or perhaps even knocking at a hotel room door, which happened to me once. This is a really dangerous game for them to play. These expatriate workers have very few Rights in the Middle East and their fate is determined at the whim of local Sheiks, Government Officials or Sponsors who hold their Passports. But such is the magnetism of Arms Cash that some take the risk. I've also been offered lifts back to my hotel by Military Officers, who during the journey are quick to move onto the Arms Cash agenda, and on one memorable occasion in the Gulf I was propositioned by my own Boss regarding an Arms Cash kickback.

Occasionally a potential Informant serving with a Foreign Embassy in London would establish direct contact in search of Arms Cash and exactly this happened to me too. A foreign Embassy Defence Attaché's Clerk called me in my UK Office and offered to release a Contract which was being held-up in return for 'certain payments' being made to him personally. This was a simple ransom demand. I didn't need to respond because his Boss in his home Country was my Company's Arms Cash Agent and the clerk found himself posted abruptly.

Executives from smaller Defence and Security companies sometimes approached NDC to offer to facilitate or broker Contracts in Countries where they felt their connection was strong or where NDC had no or ineffective representation, and independent 'Freelance' brokers also approached the Company. Such offers might come with the promise of kickbacks to encourage 'co-operation', a much-used word in the Industry. A Palestinian Broker I knew in Bahrain had a mantra which covered the subject nicely '*Prosperity through co-operation*' he used to say. It must be said that his personal Address Book, nothing digitised for obvious reasons, would have been worth a fortune, and wherever I was travelling he was sure to have a contact there; the Spider's web. Maybe he knew my mystery Afghan caller, maybe he knew the former Ambassador. Probably; it's a small world.

Some Customer Staff in Military Offices responded badly to a British presence which can still agitate the old wounds inflicted by Imperial history; a sure-fire business inhibitor in some countries, especially in Africa. I was chatting with a Somali Clerk in Abu Dhabi who told me that the British had placed social and political time bombs all over the world but especially in his home country. Arbitrary and artificial political lines drawn on maps by British Colonial Officials separated tribes, friends and families and ignored traditional rights and natural resources such as Wells and grazing land he told me.

He was well educated and highly-politicised and his knowledge demanded from his Listener a deeper more insightful historical perspective than is offered by the vacuous National Curriculum at home. It was far from a Radical's rant and I enjoyed our conversations, until he asked me to bring him a copy of Rushdie's 'Satanic Verses'. Copies were banned, but he wanted to read for himself what the controversy was all about. I warmed to his curiosity but travelling in the Region with a copy of *that* book was a jail-baited invitation. In another Middle Eastern Military Office, I chatted with an Arab Warrant Officer whom I took initially for a Syrian from his dialect, but he corrected me in his excellent English, which contrasted with my flaky Arabic and he admonished me for my mistake, adding...

"...you bastards gave away my country..."

Welcome to the Office! He was Palestinian-Lebanese of course and I asserted that I wasn't personally responsible for the theft of his homeland and could not account for those 'British inspired' events which triggered his loss; I had encounters like these all over the Region. History rankles and tortures more deeply in countries touched by Britain than is ever appreciated at home and understandably and justifiably so. It is our folly as a Nation that we have forgotten our own past, as the reaction to Gaza today makes clear yet again. Britain has had an enduringly negative effect on so many populations.

My Arab Army experience helped me a little at times like this, since I'd heard unvarnished opinions about this stuff first-hand from many Arab Officers and I shared some of their sentiments, though not others; but in true Defence Industry style I wore the Mask of Janus since I also made business trips to Tel Aviv. Thin ice Ralph, very thin! I'd once mentioned my Jerusalem visits to a Palestinian friend in London and he'd become emotional because it was his life's ambition to live there, but for a host of reasons I cannot recount here, he was unable to even visit the place and his enduring pain was immense. Milky British Politicians consistently and persistently underestimate the depth of such feelings, but it was physical feelings which distracted me now.

25

I bounced around painfully on the poorly sprung rear seat of the taxi en route to the Headquarters and wrestled mentally with a scenario far removed from my usual routine. I was now dealing with a death threat and a critical meeting which could make or break a multi-million-pound Arms deal. The taxi ride ended at the entrance to the HQ compound which housed the Procurement Offices and I unfurled myself from the seat and sat fully upright to look around.

The compound was situated on an isolated site outside the City and contained numerous single and double storied buildings, surrounded by a buff-coloured, rendered concrete-block wall about three metres high. The wall was topped with coiled razor wire and illuminated at night by a variety of angled floodlights. Elevated Sentry boxes afforded their armed occupants panoramic views over the approaches. None of the compound's interior was visible from the road and the sandy-coloured walls blended seamlessly into the desert backdrop.

Between the Compound and the public 'black-top' road was a broad dust-blown expanse of sand and stones, crisscrossed by vehicle tracks and decorated with odds and ends of rubbish; notably, empty soft-drinks cans and discarded pink and blue flimsy plastic bags, which fluttered in the rare breezes when caught-up on the spiky little plants which struggled for life there. A single stunted tree provided meagre shade, and the area beside this tree was the usual waiting point for taxis which ferried visitors to and fro.

I looked around at this dusty and exposed parking area, occupied as it was by two or three other vehicles; taxis weren't allowed through the Compound gates. Per usual practice I instructed my driver to wait for me there and I left the Cab and walked briskly across the sandy tracks towards the Compound gates. The taxi lacked air-conditioning, so I was already hot and sweaty. Exactly what you want before an important meeting; equally, once out of the Cab it was hot and sweaty anyway. Oh! face it, it was always hot and sweaty.

Armed guards gazed down disinterestedly as I approached the gateway, where tall, grey-galvanized steel double-gates hung beneath a decorative archway. A smaller side gate admitted pedestrians and western Carpetbaggers like me, and a national flag drooped limply atop the arch in the shimmering heat. Defence Company visitors could access the Compound on foot only and had to be checked-in and out through a Visitor's hatch in the Gate Guardroom; once inside the perimeter walls, the nearest administrative buildings were still some distance away. I felt very exposed as I stood outside the Guardroom, looking around and constantly scrutinising every passer-by and every passing vehicle as I waited for clearance to enter the Camp.

Only those Western Executives who've tried it themselves can even begin to understand that simply gaining access to many Middle Eastern Military bases can be a serious challenge. Soldiers manning Reception areas or Guard Rooms have absolutely no interest in your visit and might not, more probably do not, speak any English; why should they? They see no reason to respond with urgency to any approaching Westerner, in fact quite the reverse, and the more irritated and agitated an overseas Visitor might become at the length of time it takes to get through the Gate, the longer it will actually take. It's a form of negotiation. Even first-time visiting Brits' imbued with, and trained from childhood with the spirit of queuing, are likely to wait a very long time at the average Middle East Guard Room; more so, if it is not a Headquarters, where visitors are commonplace, but rather, some distant Unit Base in which City habits and protocols remain unpractised.

Unwary Executives might be overwhelmed by the sudden arrival of more Soldiers, stopping-by simply to exchange greetings with their friends inside the Guardroom, in which case the visiting Executive may be shunted backwards, and will for all practical purposes become invisible as the soldiers greet each other and catch up with the news. Soldiers in Middle Eastern Guardrooms do what most soldiers do in most military Guardrooms all over the world, except they don't look at Porn (well, they do, but not in their Guardrooms). They read newspapers or magazines, they drink tea and coffee, they chat to their mates, they smoke, they send texts, they sleep, and of course in the Middle East they stop work to pray.

Guardroom personnel are immune to urgency and they have been selected for their inability to respond to foreigners. It's their country though, so fair enough. In my own experience it's best to allow at least an extra half an hour or more to negotiate the Guardroom hurdle. Those Western Executives without any spoken Arabic should waste no time at all on this activity, but rather, should go straight to the nearest airport and fly home, where they should consider a change of career.

It was standard practice for Guardroom Staff to take Visitor's passports, which disconcertingly disappeared after a cursory comparison of photo-to-face. Sometimes though not always, there would be a badge or lanyard Pass issued in exchange for the Passport, which hardly ever seems to me to be a fair exchange. Persuading the Guard to call the right extension for the Office to be visited becomes the next challenge, even if you think you know the right number. It's reasonable, indeed sensible, to expect a few wrong number calls and a lot of social chit-chat on the line. You are not their priority. Expect a long pause wherein nothing seems to be happening; often one will be invited to sit whilst this 'nothing' happens. It's a bad sign when the coffee is offered as there's a long wait ahead. Another pause is likely until transport issues are resolved and all entry procedures are likely to be repeated on arrival at the actual destination building within the Camp itself. Executives could be forgiven for giving up.

A Defence Sales Urban-myth holds that somewhere in the vast and featureless Sweihan Military Cantonment area outside Abu Dhabi, a Western Defence salesman is doomed to drive forever around the vast and featureless area of the Base, like some demented arms-selling Flying-Dutchman, destined never to find the Office he seeks and I've narrowly avoided being that Salesman on several occasions in several countries. Although my recent visits to this particular Compound ensured actual recognition at the Guard Room, there was still a struggle to achieve a telephone connection with my Sponsor and it seemed like an eternity until I was cleared to enter. Finally, a sand-coloured minibus took me off at its own leisurely pace to the Contracts Office.

The meeting itself was almost an anti-climax after the death threat preliminaries and it lasted much longer than I anticipated: it carried me a big step closer to the deal I sought. It was a good meeting attended by senior Technical, Contracts and Operational Staffs, and it was a Power-point triumph. I made product presentations and discussed 'indicative' pricing and general contractual terms with the Contracts Officers present. I'd previously delivered deeper technical presentations to the Users, so this meeting was really a confirmatory stage. I sustained all the outward signs of contact neutrality, whilst behind the scenes I knew that battles for influence and Arms Cash were raging between various vested interests, including those of some of the Officers and Officials now in front of me. With the right connections in place these meetings, presentations and discussions would be no more than cosmetic dressing to sustain the illusion of normality.

We discussed possible schedules, so it really was beginning to look promising, superficially at least, and importantly, the Requirements about which I'd previously learned secretly seemed to be very real. There was some 'side salad' too in the form of an additional requirement for Small Arms ammunition. Normally we wouldn't bother with this low-end stuff, but it was a reasonable quantity, around twenty-five million rounds, which NDC could buy-in and wrap into the overall package; 'small beer' in the great scheme of things, but somewhere else to spread the Arms Cash costs. Without the right Agent and without Arms Cash I knew that nothing would happen here; there would be no Business and no contracts for NDC. We completed the discussions and I headed back to the Guard Room where I exchanged my Pass for my Passport, always a relief, before walking out into the dusty car parking area to collect my taxi. I looked around. The taxi had gone.

The completely empty parking zone was an acutely depressing sight, a heart-thumping moment. Why had my cab gone, where was it? I'd been very clear with the driver that I'd be some time and he was to wait for me. Nothing unusual in that, the drivers always waited. The camp was miles from the City, and from anything else for that matter, so it made sense for cab drivers to wait as they wouldn't find a return Fare here. He was going to get well paid for the waiting time, why had he gone? This was hardly Paris in peak season and the City area itself was always quiet at this hot time of the year.

Was it my imagination or it was unusually still and quiet outside the Camp? No cars drove past and there was no one in sight. It was like one of those old pre-sniping vacuums in Northern Ireland when a whole street could empty suddenly without warning prior to a shoot. The fragility of my situation recrystallized very clearly. It was a dumb idea perhaps to go to the meeting after the death-threat and a dumber idea still to be in the ever-troublesome Middle East yet again; Face it, Defence was a dumb Industry to be in.

It was a bit late for self-recrimination or regret though; useless sentiments. As ever, having limited options did not mean 'do nothing' so I made a point of presenting as difficult a target of myself as possible (the Army used to call this 'hard-targeting') whilst I thought about it. The Taxi had gone, there were no others and no cellnet connectivity. The area between the Camp and the City was not one of western-style sidewalks and shade because no one expected to walk there. Like the area outside the Base it was dusty and empty with shimmering heat for good measure.

In my lightweight business suit and carrying my laptop backpack with overnight kit stuffed inside, I hardly blended into the local landscape, in fact I stood out like dog's balls. There was no alternative other than to start walking, which I duly did, and I removed my jacket and stuffed it through the backpack straps and walked on. There, I hardly looked British at all now. Sparse oncoming traffic was heading away from the City and I was getting bemused looks from drivers. Occasionally a heavily laden fume-belching truck festooned with jangly Asian decorative chains and driven by a turbaned Bangladeshi chugged past. I sweated in the heat and sand quickly infiltrated my useless urban shoes to make walking across the lumpy ground even more uncomfortable. I felt like shit.

As I plodded along sweatily I completely missed the approach of the smart black Pick-up truck which drove up from behind and slewed to a halt beside me, kicking up sand as it stopped. A tinted electric window zooshed down and the Arab driver looked out at me from his cool air-conditioned cab. I froze momentarily half-expecting him to produce a gun. He didn't of course.

"Going to the City…?" asked the dish-dash clad driver in good English, pointing towards the City for emphasis. He offered me a lift but looked bemused when I hesitated.

"It's OK…" he said smiling "…I'm a Police Officer…" he laughed and reached up to the sun visor and pulled out a Police ID Badge which he waved at me through the open window. I couldn't read the words, but the lure of the cooled-air interior was irresistible, so I chanced it and clambered into the cab alongside him. We set off towards the City.

It was just a bizarre coincidence completely disconnected from the early morning death threat. This Police Officer, Mohammed of course, asked me where in the City I was going and why I was walking, where was my car? Nobody walked anywhere. I explained that my taxi left without me and he laughed. I complimented him on his English; where had he learned to speak it? Some coincidence indeed, he was a Police Firearms Trainer and before his Instructor-training in the UK he'd been sent to Bournemouth for an English language Course. We spent the rest of the journey discussing handguns and ammunition and he dropped me as I requested in the City Centre, where I hailed a taxi and set off to one of the smaller more obscure hotels I knew, determined not to make myself available for being snatched from the Airport or from anywhere else for that matter.

I checked into my new hotel and told the Indian Manager I was taking a break for a day or two and didn't want to be disturbed. It was a relief to get out of my sweat-soaked clothes and into the shower. I'd yet to make any progress with the threat situation and hadn't contacted either the Embassy or NDC. In the case of the latter, I didn't believe the Company either could or would do anything about it, and anyway I didn't want them trampling all over the place messing things up.

With fragile network connections, reliable comms could only be achieved over a landline or ageing Fax system, both of which I treated as insecure. I managed to voice-call a friend of mine and using military jargon I explained my situation, just so that someone else at least knew there was a problem. It wasn't the first time that the combination of army jargon and military radio voice procedure had come in useful. It was quick and it baffled unfamiliar listeners.

I stayed inside my room taking meals via Room-Service, such as it was, and tensing every time I heard a noise outside the door. The Room Service trolley squeaked dreadfully; I thought an Oil-rich State wouldn't have such a problem. The hotel was almost empty so there were periods of complete quiet, save for the hum of the air conditioning and vibrating rumble of traffic outside, which shook the poorly fitted double-glazed windows; this grubby little hotel was a far cry from my usual haunts but it would do well enough in the circumstances. The pre-weekend exodus from the City gathered pace and although fatigued, I slept fitfully.

The following morning, I set off for the Embassy, making the journey in two separate legs and using different taxis. I walked back from the second drop-off point to scan the area. Today, many British Embassies look more like defended bunkers than Embassies, with high perimeter walls, barbed wire, CCTV systems and entry-control equipment. Some have become difficult places to access even for British citizens and can be forbidding places to visit, which is the price to be paid for foreign interventions. 'Democracy' buys you self-imprisonment and ironically makes you inaccessible.

I didn't really expect the Embassy Staff to be of much help and I wasn't disappointed. Reporting the incident required care about which details I gave Officials. I was hardly going to say '...some mad ex-Ambassador wants to kill me because I won't sign him up for a million pounds worth of illegal Arms Cash' but I felt the need to at least pinpoint the Ambassador as a probable source of the death threat, so I made up a story about a trading dispute. I just wanted them to log his name in case anything came of it.

The Embassy Officials sympathised weakly but couldn't offer any practical help and I hadn't expected more. I told them I just wanted them to log the incident and specifically the name in case anything happened and this they did. They weren't interested. I was disturbing their steady expat weekend routine, always an error. I walked, then taxied indirectly back to my hotel, doubling back a couple of times for good measure and I also recced another small hotel to which I could move after another day or so. Once safely back in my room I played one of the few cards I had left and made a local call to Abdullah, a friend and business contact I had in the City.

Abdullah and I had established a good rapport working on small Security Projects when I'd represented a different Company a few years earlier. He was well respected in local commercial Security circles rather than in the hardcore Defence domain. I didn't explain much over the phone, I told him I was in-country and simply wanted to meet up. He came over to the hotel that evening for a meal and I explained (a version) of what had happened. He was angry about the threat and left promising to do something about it, saying he'd be in contact the next day. Another tense night passed with yet another curry-flavoured room-service meal.

Abdullah called me the next morning then came over to the hotel as he didn't want to discuss anything on the 'phone. He told me that actions were in hand to put things right and I was to be ready at midnight for a vehicle pick-up from my Hotel. He would be with the car so that I knew it would be OK. It was a stressful wait and despite misgivings I was ready in the small hotel lobby at the appointed hour. A smart-looking large black Mercedes (as ever) arrived punctually outside the hotel and I watched from the lobby window as Abdullah, who was in the passenger seat, jumped out of the car to hold open a rear door. He looked expectantly towards the hotel, so I bustled out of the building and jumped into the back of the car. Abdullah climbed in beside me.

We were driven away from the City Centre and out into the night by an unusually competent driver; I recognised some of those driving techniques. We travelled on in silence and once clear of the urban landscape with its well-defined and identifiable buildings and streets, I became a little disorientated, this wasn't a route I'd used before. Unsurprisingly the landscape was predominantly sand and stone, cut through by shallow wadis and modulated by low hills.

The ever-brilliant Middle Eastern night sky with a bold moon provided helpful illumination, but the stark and rolling landscape offered few visual reference points. With no 'phone GPS available I calibrated direction from the stars and checked the time to estimate distance. We headed North and drove on for about thirty minutes averaging around 70-90 kph I guessed, until finally we left the 'black-top' and drove along a dusty track (aren't they all dusty?) to a large compound surrounded by a crenelated battlement wall; something from Beau Geste or Hollywood perhaps. A large pair of ornate gloss-green painted steel gates set into the high sand-coloured walls and flanked by palms, opened before us as the car approached and closed behind us after we swished through. I felt as if I'd drifted onto a film set.

Within the walls and extensively illuminated, was an attractive formal garden complete with irrigation channels like an Omani Falaj, and a Well fringed by Date palms and lush vegetation. The car halted in front of a Colonial-styled but modern construction Villa, which was coated in the standard sand-coloured render, and which displayed a national flag which hung uselessly, like all the others, from its roof-mounted flagstaff. There was nothing unusual about the presence of the flag, which did not in itself signify a special residence, the city was full of National Flags most of the time. Everyone had one it seemed.

As we approached the grand steps which rose to the house, a smart dish-dash clad servant, Bangladeshi probably, appeared at the grandiose wooden Indian-style front door and gestured to us to enter the building. He bowed slightly as we entered. We were shown into, and waited within, a classically cool marbled foyer of double-storey height, with a central ornamental fountain bubbling away to itself. For a few minutes all was still and quiet and then I heard a voice.

"A salaam aleykum! Good evening Ralph..." I turned in the direction of the greeting to meet my new friend, a Royal Sheikh, and of course the Head of Defence Procurement. We'd met previously but formally at the Military Headquarters on previous business visits and I glanced quizzically at Abdullah who simply shrugged his shoulders and smiled back at me.

My host was a shortish, stocky man, not old, perhaps forty. He wore a crisp, plain, pristinely white cotton dish-dash with Kimma and was clearly amused by my surprise at this piece of theatre. He led us into a large Reception room, furnished expensively in the regional Arab style where he opened a large bottle of Black Label Whisky; very welcome too in this officially 'dry' State. He poured generous shots for each of us and smiled broadly at me as he spoke.

"I'm sorry to hear that you have had a few, er, difficulties..." he said "...don't be concerned; these will be resolved..." It was a very simple statement with the twin elements of chill and certainty about it. He had 'presence' about him and the aura of one used to a position of power and influence; one used to instructing rather than asking, one used to getting what he wanted. He was courteous, well-spoken and I'm certain, completely ruthless. All of this was apparent within minutes.

"...you were right to be cautious..." he continued "...the problem was serious, it was authentic. We are addressing this matter...."

The three of us spent an entertaining night drinking his Whisky, which certainly eased the pressure, and oddly, a Chinese meal was suddenly produced as if from nowhere and served by the Sheikh's Staff from those foil containers in which take-away food is sold. It was a European-style meal in the conversational respect, since in more traditional or formal Arab society we might have eaten in silence to respect the food. We discussed neither the death threat nor the ex-Ambassador and we didn't discuss any Defence business either, but we did share an illuminating geo-political discussion which yet again recalibrated my own regional perspective.

British political opinions concerning many world regions, particularly the Middle East, are still shaped by ageing post-colonial attitudes which have failed to keep pace with regional evolution. That's why the Brits keep making mistakes overseas. It's a major failure which reflects the unengaging, hunkered-down posture of so many Senior Diplomats. That the British Foreign Office 'take' on regional matters was a long way wide of the mark, was a recurrent theme during most of my international travels, and made me wonder if our Embassy Staff ever actually met any local people; probably not it seemed, certainly, none that mattered.

After a couple of hours with my new friend it became apparent that it was time to leave. The Sheikh bade me farewell in the marble foyer and advised me once again that the '…issues would be resolved' adding that '…we will find a way to work together…' It wouldn't be appropriate for him to meet me openly so my city friend and fellow Black Label drinker Abdullah would become the conduit for communication. The Sheikh and I shook hands and Abdullah and I returned to the car. The green gates opened to release us into the cool early morning world, where Omar Khayyam's 'Hunter of the East' had just '…caught the Sultan's turret in a noose of light.' Quite a morning so far.

The car sped off with me feeling partly reassured but certainly a bit confused too. One certainty I did have was that I now had a channel to the Head of Defence Procurement, even if the circumstances which led me there had been asymmetrical to say the least. I'd also been able to plunge the threat scenario into a glass of Black Label, so that was some consolation too. Only now, years later do I recognise just how desensitized I'd become, as my focus then flipped so easily from managing or evading a personal death threat to 'running' potential contract revenue numbers through my head; it was a pleasant enough distraction though, the numbers were huge after all.

When I quizzed Abdullah about the surprise meeting, he explained that he shared some non-Defence business interests with the Sheikh, and they enjoyed a personal friendship. Abdullah fulfilled a discreet role as the Sheikh's 'eyes and ears' in local Business circles, keeping tabs especially on anyone critical of the Royal Family-led Government. These folks were not distracted by Democracy, as that Western political obsession doesn't work for everyone. Citizens in this State might not have voted for their Leader, but look on the bright side, they didn't pay Income Tax and they enjoyed world-leading healthcare for free.

I hadn't known previously about Abdullah's connection to the Sheikh, so it would never have occurred to me to ask him to connect us. Maybe he could have fixed a meeting ages ago, and that would have been a lot less stressful than receiving a death threat. But my previous

35

business activities didn't merit such a high-level contact, hence I guess we'd never discussed the Sheikh before.

Abdullah explained that following my call, he'd briefed the Sheikh about the death threat, and the Sheikh insisted on meeting me personally at his out-of-town residence, not simply as a favour to me, but because he had his own agenda and was focussed larger prey. He knew what was coming down the line in the form of future National Defence needs, so teaming with NDC could be a marriage made in heaven for him; the vows were there waiting to be exchanged, and Arms Cash dowries were available. It became blindingly obvious that the Sheikh's influence was much greater than I'd previously imagined and was not restricted to his day job as Head of Defence Procurement. He was truly a leading inner Royal-circle figure.

Abdullah also told me that the ex-Ambassador had been confirmed as the source of the death threat and that it had real substance, insofar as arrangements had been made for my abduction. The Sheikh was hugely displeased, mainly because large-scale Defence business was his preserve, and the ex-Ambassador had trodden upon his Royal toes, which was a much more serious transgression than threatening to kill a disposable Defence Company Executive. The actual Caller's identity eluded us all at the time, although it seemed to matter less and less. In what remained of that early morning, I didn't 'sleep the sleep of angels,' but I slept well enough, confident that something was going to happen to alleviate the situation, even if I could not be completely sure as to what that 'something' might be; I'd have to take that on trust. Trust? In the Defence Industry? In this Region? Really?

Chapter 2
Follow the money

The following day nothing happened, so I sustained my room-bound low-profile routine in a state of tense anticipation. Abdullah finally returned to the hotel the following morning and took me to a very smart City-Centre Office Tower building. The umpteenth-floor Office interior was impressively swish and plush in the modern western style, and it was there, this time in Abdullah's company, that I once again met up with the threatening former 'Ambassador Extraordinary and Plenipotentiary'. The ex-Ambassador was ushered into the large Office suite by two tough-looking Arab Minders and his demeanour had changed remarkably.

Gone was that assured, arrogant, and demanding swagger which so characterized our previous meetings, replaced now by an almost supine, apologetic figure who craved forgiveness for what he'd done. He'd made the threat against me, yes, and sought someone to carry it out for him, which is how my mysterious caller came to know all about it. Someone had leaked it out. I never caught up with my Afghan caller to thank him in person although I did eventually identify him. No one knew where he'd gone after leaving Singapore from where he'd made the warning call. After his apology, the Ambassador was escorted robustly from the Office, quivering and trembling, and I neither saw him nor heard of him ever again, despite numerous visits to the Country. I never heard anyone else there even mention his name either; it was as if he'd ceased to be. H'mm.

I was relieved to be free of the threat, but part of me recognised in the Ambassador's new condition a disturbing and unsettling image which sent its own message. What pressures had been brought to bear on the man and what consequences had he faced? If they could do it to him, they might do it to me one day. But earthy reality kicked in, what did it matter to me what happened to the Ambassador? He'd threatened to have me killed so he deserved everything that came his way, and at the time, I'd happily have double-tapped him myself.

Now the Ambassador had been shown the door and the Sheikh and I had met up, things could be properly organized. I sent NDC a coded message and gained approval for an Agency Appointment, which would have to be formalised in a neutral location between

NDC's shadowy Agency Director, Harry Kensington, and the Sheikh personally. It would be just the two of them, maybe with a Lawyer each, and the single original copy of the Agreement would be held in a neutral location; Switzerland maybe, or somewhere in the Caribbean, or these days in encrypted 'Cloud' form. Both parties would have to be present to access the document to make any changes to it, 'dual key operation'. Such were NDC's and indeed the Sheikh's operating protocols. We codenamed him 'Hector'.

As was usual in Agency arrangements, no third parties had any rights of decision on behalf of Hector, and nothing was delegated in the Agreement. Abdullah and the Indian gofer Bandar were simply conduits for Hector, local cut-outs used to relay information and plans. I briefed Hector on the NDC appointment process and requested a further meeting to discuss and settle some of the essential details. He was clearly very familiar and comfortable with this sort of thing. The new Agency Agreement would be prepared in NDC's London Office prior to the clandestine signature meeting somewhere, but I would take no part in that, as the appointment process was Harry Kensington's exclusive province.

Hector and I haggled over Commission rates and several different rates were agreed, which is pretty much standard practice, for it was not always the case that a single rate of Commission applied to an Agency Agreement. There were relatively low rates for materiel which was proprietary to NDC. Indeed, it could be argued that NDC shouldn't have to pay for proprietary systems at all, although in reality, Hector had the power to change an entire Inventory if he chose to, and he could put NDC out of the running forever, so it was akin to paying a Licence Fee to cover the range. An enhanced rate was agreed for business I wanted to take from a Competitor which Hector already represented, and this rate was a one-off on a small part of the deal at a big fifteen percent. This was much higher than usual but reflected Hector's status and his change of allegiance in NDC's favour in this single case; he would not have switched that Agency for any less.

This high rate didn't matter a jot to NDC because the Corporation would benefit for years to come from high-value contracts. Indeed, Middle Eastern customer pricing on NDC Systems generally was set at around 300% above European levels. Arms Cash Rates for Mainstream systems were generally worth around five to ten percent commission, and some other unsexy stuff, spares, etc., five percent only. When he met up with Kensington, Hector would have to agree a different commission schedule for any of NDC's super-sexy systems, which were outside my Brief. We agreed enough to get everything started. Abdullah would be Hector's link during my in-country visits to reduce Hector's exposure risks and his Indian gofer Bandar would be the Defence data courier.

Hector had everything compartmentalised neatly with different people handling different things for him, so only he ever had the full picture. Hector would take care of Abdullah and Bandar financially, and NDC had to take care of him if the Company wanted any contracts. I knew that 'NDC Corporate' would take over and run Hector directly from the centre to meet their strategic objectives, and that I represented just the booster-rocket section in the process, to be jettisoned when burned out; only 'Corporate' got to go to the Moon with Hector. I would be reassigned to start another fire somewhere else.

I changed Hotels again because I now needed somewhere bigger and more sophisticated with good Office facilities and proper Restaurants. I'd peaked on Room-service Curries. I started work immediately on formal Proposals for submission 'through the front door' to this Defence Customer. Time was running very short now as Ramadan was even closer, and this activity wouldn't usually have commenced until after the Agency Agreement was signed, on the usual Defence Industry basis that no one Party ever really trusts any of the others. But Hector and I had to trust each other because Programme time was very short and stuff had to be done.

If NDC changed their minds and backed out of the Agreement at this stage, I personally would be toast, and NDC could kiss goodbye to any future business in this State, a fact which Sir CEO would have to explain to the Foreign Office and the MoD, both of which were well aware of and supported NDC's previous strategic, though naïve, in-country business agenda. Without Hector in tow, NDC's aspirations were just wet dreams. For his part, Hector ran the risk of personal exposure, which would be regional Political dynamite, so we were all locked into the Industry's most effective unspoken relationship. Exposure risk remains the critical bond for such arrangements; loyalty was never a factor it was all about the Arms Cash.

We, mainly me that is, worked in earnest on Proposals. Bandar brought me key data on USB sticks and I recalculated and reprogrammed aspects of the Country's Defence holdings to suit NDC's portfolio and prepared operational data for Hector to use to demonstrate to his Staff the credibility of the figures. I also wrote discreet business case notes for him to support the selection of NDC equipment to 'justify' the purchases. I was exchanging data at a furious rate with NDC Commercial Departments in the UK, and this process was both complicated and stressful.

Pricing activity always caused friction between Sales, Marketing, and Commercial Departments, 'Never-Never Land' as they were known. None of the Commercial team would know Hector's identity, and they resented being told how to formulate the Offers and what to add into the pricing matrix for 'Corporate Marketing', i.e., Arms Cash costs. I was well-practised in the art of on-the-hoof pricing because I'd worked internationally for smaller Defence Companies, which both needed and empowered me to price locally to win business for them, so I enjoyed an earthy frontline familiarity with market prices which the Commercialists lacked. I needed them to be quick with their responses, but that wasn't how they liked to work.

I also needed to get real documents in front of the Customer in quick-time and could ill-afford Corporate turf wars. If nothing else, my Middle Eastern Customer had every realistic expectation of having to use his War materiel, whereas much Defence Export procurement was made simply to reap Arms Cash rewards or to satisfy Ruler vanity. So, time was of the essence and I couldn't afford the luxury of following the umpteen step, disabling, Corporate 'Bid Process' which had somehow become mandatory, and to which NDC's Commercial Departments were incurably addicted. They no longer managed risks they ran from them.

Another complication was that Divisional costs and margins varied across NDC's empire, depending on the nature of the Systems or products, but it was important for us to aggregate figures across our Proposals to 'lose' the different Arms Cash percentages. It all needed to look smooth and spike-free to the Customer to avoid attracting undue attention. Even with a top Agent in place, priced proposals need to look as 'right' as possible with no spikes.

Viewed as an overall package, NDC's profits would be simply huge, but the individual Divisional Directors didn't see it this way, because each Division was set-up as a profit-centre, a mini kingdom. When these Directors couldn't realise their desired level of profit from an Order to boost their personal bonuses, they would not bid. Hence more nuanced calculations and manoeuvring for me. On my own and far away, I couldn't overcome these concerns over personal bonuses, for which their appetites were insatiable, and which trumped any pan-Divisional co-operation for the 'Corporate good'. I needed someone back at the Ranch to bang their heads together.

I slaved away in my hotel room, taking out time to confer with Abdullah and Bandar and preparing and delivering new Customer Presentations to keep-up appearances. The whole process was made more stressful by the time difference with UK which ensured that sleep was in short supply. NDC's Divisions whined and whinged about it all and left to their own devices it would have taken them months to generate the Quotes, if ever, and probably given a free choice they would have withdrawn. In their world it was much easier simply to sit back and wait for the next fat UK MoD order to fall on the mat with no effort whatsoever required on their behalf. It remained a mystery to me as to how NDC ever managed to issue any Proposals.

Risk-averse Commercial Departments have become the tails which wag the Corporate dogs these days, with the result that billions of pounds worth of potential business is lost regularly by UK-based Defence companies. I've personally lost numerous good potential Projects because of 'chicken' Commercial decisions. Risk-aversion has become a religion and its greatest devotees work in Defence Company Commercial departments.

NDC's commercial ethos, like most UK-based Defence Companies these days, called for risk-free solutions in every respect and a cloying safety-first culture dominated their world to the near exclusion of all else. Commercial Staff distrusted and invariably disliked Sales and Marketing Executives, these two categories being chalk and cheese psychologically. We didn't have months to play with we had only days, so I was hot linked directly to the Corporate Commercial Director to get it all done, which made Divisional Directors whinge and whine even louder. They hated me, but that was nothing new.

Most major UK-based Defence Companies have become self-led not Customer-led; they are introverted not outgoing and inflexible processes dominate their behaviour. In fact, they have become slaves to 'process', that killer of opportunity. They are certainly not slaves to Winning. Customers were expected to fit around Company processes and conditions, rather than the other way around, and Sales and Marketing Executives were enemies of the State.

I was well placed to achieve significantly better margins than any of the Divisional Commercial teams could achieve, as they did not have the real-time feed which I was receiving on an almost hourly basis, and of course they couldn't be connected to it. It wasn't practical for me to shuffle between the UK and the Middle East simply to brief Departments and massage egos, or to update their internal figures, that was way too slow. It all had to be done now because we had a limited time-window in which to achieve Contracts and did not have the luxury of endless internal Bid meetings to wax lyrical about it all.

The real complexity, the sensitivity, was that I knew before we quoted that we were going to win but the Commercial Departments at home did not. They were not privy to the scripted activity and couldn't be made aware of NDC's 'arrangements'. They didn't know how this Contract Win was going to work, they just had to add-in the percentages they were told to add in. Left to their own devices they would either have priced products too low, as if they really were in an actual competition with much leaner Competitors, or they would have gone too high by over-assessing commercial risks and adding fat contingencies; that was the most common practise and usually resulted in failure.

The reality was that Competitors had been eliminated from consideration before the official 'Requirement' was even issued. Some would be adjudged 'non-compliant' as in some UK domestic MoD bidding wars, whilst others might be deemed unable to manage the volume of supply etc. Others might be classified as financially unstable or technically incapable etc. In this respect it differed little from the UK MoD process of issuing Pre-Qualifying Questionnaires (PQQs) in advance of Tenders. The questions and the 'thrust' of any PQQ can be skewed and adjusted to favour some Companies and to eliminate others, and I prepared data for Hector to justify selecting NDC in just the same way.

NDC's Divisions and I were at war over the whole thing and internal battles raged at home, fuelled predominantly by Executive egos and the hypocrisy of Corporate life, i.e. Divisional Directors outwardly embraced corporate 'team-player' ethos, until that is his or her own Division was competing with others. Then it was everyone for themselves. It was all a bit pathetic really, as Customer focus was the last item on their Agenda. These handbag-hurling conflicts 'twixt Sales, Marketing, Bid and Commercial Departments, reminded me of the equivalent military clashes between Intelligence and Operational Staffs. It's the simple difference between those who actually have to do the job on the ground 'hands-on' and those who pontificate about it and don't.

The Arms Cash component in any deal and the resultant need for total security for the Agent always created conflicts at home. You simply can't tell everyone what's going on and why certain things had to happen in certain ways at certain times. You just need everyone to do as you request and not ask loads of difficult or stupid questions, especially the latter, which predominate. The selection of Arms Cash Agents also created friction when Regional Salesmen were also not privy to the secret arrangements made by their Corporate centres. Divisional Salesmen or Business Developers will be quick to blame their Corporate Marketing Department counterparts for appointing the wrong people, when Sales are difficult to make, i.e. 'We didn't get the sale because we had the wrong Agent'.

I worked into the small hours every morning with my calculator burning as hotly as my ashtray; I don't think I'd have got by then without cigarettes, but I can't stand them now. I made more token 'through-the-front-door' visits to the Headquarters to make presentations and to submit priced or unsolicited Proposals and this was normal enough during any in-Country visit and attracted no undue attention. It would have looked odd if I had I not done so. Despite my newly established and reassuring 'relationship' with Hector, I maintained an irregular and low-profile routine, and it would have been an understatement to say I didn't go out much. Sad bastard! Why did I even do this work?

The final output of all this activity was a series of nicely 'pre-tuned' Offers which Hector and I had structured discretely, and which were sent to the Customer directly from NDC in UK. The whole performance gave the appearance of a fully open and transparent Bid evaluation and Proposal process, which resulted in contract wins for NDC. In reality, it was a total 'construct' arranged between NDC (me) and Hector. I relied on my boss, NDC's Corporate Marketing Director, to keep Executives at home in-line as all this progressed, because they were not 'in' on the arrangements. I imagined a heavy lunching programme in London as the mechanism for this inter-

43

Directorial activity. But this was not kid's stuff. The implications of getting caught-out were severe, but if we hadn't been doing all this then one or more of our Competitors would have been doing it instead, and there would be no contracts. NDC was lucky to make the connection first, and apart from the revenue, success could be measured in the number of jobs sustained and created at home. It's easy to be pious about Arms Cash, but it's not easy to win big international contracts and keep people in jobs. Having said that, it must be said that I'd some time ago lost my 'Selling for GB Inc.' delusion.

Most leading UK-based Defence Companies, and I've worked for many of these so I can make comparisons, still have a problem with the 'Sales thing'. Sales is still not an honourable profession in UK, it's all a bit grubby, still something associated stereotypically with the world of Used-Cars and something to be avoided, an occupation of last resort perhaps. Maybe that's why so many Defence Companies used to offer and pay Commissions, Arms Cash, to get Sales. They were palpably useless at selling anything and hated doing it. 'Marketing' was always perceived as something more skilled, more sophisticated and more socially acceptable than 'Sales', although in the Defence world not many Executives seem to understand what Marketing really is.

Most multi-Divisional international Defence Plcs had some form of central Marketing organization which oversaw regions of the world, and some of them sought out Arms Cash Agents to facilitate business. As various international legislative increments tightened, Corporate anti-corruption, compliance and counter-risk measures became, cosmetically at least, commonplace, where previously they did not exist to such an extent. In many Companies today the pendulum has swung so far that Executives are too scared to do virtually anything overseas. This doesn't necessarily mean that the Arms Cash game is over, no, far from it, it's just that for some the mechanisms and strategies have changed. For many overseas Suppliers nothing ever changed, and it's always been Arms Cash business as usual.

In a brave new digital world, Agency Agreements no longer need to be held in hard-copy in discreet locations as they used to, replaced now by virtual or cloud-based arrangements, so literally, nothing to see here. Arms Cash rewards, where they are still operated as such, no longer need to be made as old-fashioned traceable cash transfers;

actual Cash is yesterday's 'Arms Cash'. National and international penalties for contravention are now severe but it wasn't always so, and some Countries are still not handicapped by any national legislation, so the Playing Field remains uneven.

Some Western Defence Companies caught red-handed in 'Arms Cash' cases have found ways to settle out-of-court or to plea-bargain on lesser charges, a favourite being 'False Accounting' in exchange for cash fines, and this practice simply highlights their desperation for absolute secrecy. One might well ask what is it that is so sensitive to a Defence Company as to make it worth paying multi-millions of dollars in plea-bargain fines? What else could or would come out into the public domain if such legal escape routes were not available? Why do Governments even allow illegal activity by Defence companies to be concealed by plea-bargains or out-of-Court settlements, what do *they* fear? All of this is high-level stuff though, compared to the mundane but frictional relationships between Defence Company Departments at ground level, and working-level Sales Execs like me.

When I first worked internationally, I coordinated my Sales activities through a Corporate Regional Marketing Executive, usually based overseas, who in turn controlled the Company's Agents in 'my' territories. Not knowing what was going on behind the scenes was frustrating to me and time consuming, because positioning deals took too long and genuine opportunities were missed. I'd have to brief the Regional Executive on what I needed, and he'd then confer secretly with his Agent who chatted in-turn to his 'Insider' friends (if he actually had any) until finally a version of the data flowed back to me for action.

When I later became the Regional Marketing Executive responsible for identifying, selecting and managing Agents for the Company, I too was unable to share Agent's identities with front-line Sales Personnel, so I freely confess to having been a complete hypocrite, since I've argued vigorously from both perspectives regarding the levels of information which can, or which should be made available regarding Agents, their points of contact and their inputs.

Incompetent corporate Marketeers, and oh! boy there are so very many of them, shroud themselves in the mists of secrecy as protective mechanisms against personal failure, whilst incompetent Salesmen can, and often do blame their assigned Marketeers for their personal shortcomings, e.g. '...we had the wrong guy...' This bickering interaction between Departments became comical, almost farcical at times, and overseas Customers looked on bemused when intra-company squabbles surfaced in front of them. Customers were quick to exploit these internal fissures. Coupled to this dysfunctionality, many home-based Executives, especially in Bid Teams, liked to make winning overseas contracts as difficult as possible by constantly finding or inventing reasons for not bidding. So, thanks to that heady cocktail of legislation, risk-aversion and lack of worldliness, today's UK-based Defence Industry has grown a whole generation of Business Prevention Experts, the Castrati, and it's a complete wonder that any export business is won at all.

As I established the new relationship with Hector, I confronted and managed exactly this mishmash of internal politics and limited ambition, so I needed to focus hard to get a substantial deal on the table in a very compressed time span. Hector was a willing Partner, but make no mistake about it, if a Company doesn't perform, their Agent will drop it like a hot brick; it's less so the other way around. I continued to race between formal meetings in the Headquarters and my hotel room, where I was re-programming elements of the War Stock Inventory, but my Bid Managers back in the UK were not happy bunnies. An Outsider might look-on completely bemused by all this feckless in-fighting and wonder why we couldn't all just get focussed on securing the deal, but in Defence Industry Corporations that is an unanswerable question.

NDC's Main Board Marketing Director quite rightly kept Hector under the tightest of tight wraps, for Hector was of strategic importance to the Company. He also took over co-ordination of my communications into the Divisions which helped me enormously, as the Divisions had to do what they were told by him. I was making more Corporate enemies on a daily basis and Vendettas were accruing. I might have been fighting for major new contracts in the Middle East, but Directors at home were fighting for individual status, bragging-rights and bonuses; it was yet another reason never to work for a major Defence Company ever again, so I promised myself I'd pick a small one next time, or just go freelance.

I was curious though as to why a Corporation as big as NDC had not already got a high-level Agent on the Books here in this so-called strategic target territory. With NDC's substantial marketing resources and significant influence throughout the region, it would have been normal practice for the Company to have at least one, and probably a set of high-level Agents in place here. It was odd that they'd appointed no one, at least, no one I knew about, and it didn't fit with NDC's usual pattern. But now that I'd opened the Hector channel, I'd stumbled into the big-boys playground and I was about to share one of NDC's big-boy's secrets.

I'd already bet myself that there was Corporate dirty washing hidden in this State somewhere, and I was to be proved right, although the nature of it was deeply buried, like so much else about NDC. A casual remark by Hector during one of our meetings triggered my curiosity, and other pieces of the dirty-laundry puzzle were supplied unwittingly by my new corporate 'Minder' Standish, an NDC inner-circle Corporate veteran who'd been appointed hurriedly in London to keep a weather eye on things, and to coordinate support for my exploitation of this newly tapped potential.

Standish was 'old-school', a retired Colonel, a 24-carat snob, and heaven help me, a former Defence Attaché. Triple groan. He'd been with NDC in various Middle Eastern roles for years and was a true creepy, slimy corporate animal, a consummate networker who'd clung on through the Company's many name and posture changes and purges. He was a true survivor and he was corrupt; I would have to watch my back. Through my various discussions with Hector and Bandar, and with unwitting input from Standish, I finally sussed the reason for NDC's close interest in my activity. In an unusually candid moment, Standish himself later confirmed the story to me, but only because he'd realised that I'd worked it out for myself.

Some years previously, NDC's own Sir CEO had been caught red-handed in this territory with his fingers in the Arms Cash till; he took a kickback. Coincidentally, no doubt, Standish had been the CEO's personal 'Chief of Staff' at the time of this incident and was at the epicentre of the ensuing internal corporate hush-operation, mounted to limit the damage; the Old-Boy network went into overdrive, so it never hit the Press. NDC had been desperate to keep the lid on the story as the impact would have been shattering, with severe and enduring repercussions. Was this why Standish had been selected as my Corporate sheepdog I wondered? NDC Execs used to mutter about there being a tight little 'Golden Circle' at the top of the Corporation; what made it so golden?

I spent a further week in-country completing the submission of Proposals and Technical specifications. Another week, was it really just a week? It was exhausting, with long days made longer by the innumerable late-night calls from NDC. I remained very wary of Standish, as even in the dysfunctional world of Arms Cash, there was something especially malevolent about him. I finally reached a point where I could achieve no more in-country for the time being, and thankfully I booked a flight home, anticipating a quick return to the Middle East.

It had been my turn to have the kids when I got home, but I had to 'pass' because I was mentally and physically exhausted and wouldn't have done a very good job for them; they deserved a lot better. I think they'd probably given up on me anyway by then. My head was still spinning with an admixture of relief at evading the death threat and the euphoria I felt at getting so close to a major deal. Now I had to deliver a hot-debrief to NDC's Corporate Marketing Director in London, and that was likely to be the most stressful experience of all, because I never really knew what his agenda was, even though I could have guessed.

I reported for de-briefing, having firstly dealt with the assault of the admin Apparatchiks. How many corporate protocols and procedures had I contravened this time? NDC's team of specially institutionalised Managers, the STASI, never ventured past Hastings. They never met Customers or Agents. They were the lucky ones who made it home by six-thirty every evening and enjoyed weekends with their families. These happy souls were unencumbered by Islamic working practices, international time differences, Agency complexities, or by early morning death threats, and they fully expected to still be working for NDC when Pension time came around, whereas I would either be fired or dead by then.

I couldn't reconcile their insatiable needs to account for the most trivial costs; a few pounds for an unreceipted Trishaw ride in Colombo once, a cold drink in Karachi airport on flight diversion, with the benefits of a multi-million-pound contract oozing with profit. It was impossible to convey to these unworldly dolts, that Defence Business was not actually some wonderful game played to a set of well-established rules to which all the participants subscribed. They delighted in their search for breaches of Expenses or bookings protocols, even as I positioned the multi-million-pound deals which paid their generous salaries. It's just Business, yes, I suppose it is, but

48

an Indian acquaintance has this saying '...better to live a single day as a Tiger than a hundred years as a sheep' and NDC had more than their share of sheep, whole flocks of them in fact.

Once these admin' hurdles were cleared, I went one-to-one with NDC's florid and rotund Corporate Marketing Director, the keeper of so many grubby Corporate Grail secrets, George 'Two-Lunches' East. George East headed up NDC's global marketing empire and dwelt in the rarefied atmosphere of the Corporate Board as Sir CEO's right-hand man. Blood-brothers in Arms. Here was someone who knew a secret or two, or three, or four, but he was an old-Buffer type. I was unconvinced that he'd even realised that the Berlin Wall had come down or that digitization had arrived. I'd like to describe George East as a Marketing Colossus, but in truth, he was simply a Colossus, and no Lunch was safe in his presence.

It was always best and safest to assume that I was being recorded in these cosy little one-to-one London meetings, and the real nitty-gritty of the de-brief, the Agency stuff, soon got underway. I recounted to him the events which led up to the death threat warning in my hotel room, in fact, the whole gory story of the Ambassador Extraordinary and Plenipotentiary, well, most of it anyway, for in this business, you have to keep something in your back pocket. He feigned astonishment.

"Oh, my dear chap..." he exclaimed, "...how exciting...!" Yes, he really did talk like that. Old, old School. He was palpably useless in the modern era, a throwback who dined for England at every opportunity but who manipulated corporate politics expertly; a member of NDC's Golden Circle. I'd no respect for him, but he had to be treated carefully because aside from being a consummate Corporate survivor like Standish, he was also Sir CEO's mate, and would happily 'drop' anyone who annoyed him. On the plus side, he'd managed to hold off the NDC Commercial gang from lynching me whilst I worked through my Proposals in-country, so I was grateful for that.

George East knew better than anyone that here at last was access to a high-level Influencer-Agent, where NDC previously had none, who was well-positioned to move strategic mountains for the Company, which had so much bigger fish to fry than even the major contracts I'd been fixing. It was East who'd allocated my Corporate Minder Standish to 'support me' (a very mixed blessing) and help to exploit NDC's wider Corporate potential. You had to watch these people; they were joined at the hip. In one part of the Corporation,

49

there was a very strong Masonic element, and whenever I saw these two together, I couldn't help but wonder if that was their connection, or if it was just Arms Cash which bound them. Probably both, I concluded. Maybe I should practise my secret handshakes.

NDC's bigger, sexier Tech' Divisions looked down on my Land Systems Division as a bit of a grimy working-class sort of an outfit, which represented a grubby counterpoint to NDC's glossy offerings. Practically though, the gestation periods for deals in Land Division were much shorter than their big-sister Divisions, and in a cash-burning inferno the size of NDC, quick and regular cash-infusions were needed to pay Executive bonuses and Arms Cash bills, to say nothing of the Marketing Department's travel bill, in itself, equivalent to the GDP of a developing Nation.

NDC's Tech' Systems Executives could and did spend years promoting and managing single major campaigns, without ever having to actually sell anything, whereas Land Systems business cycles were shorter and the results, or lack of them, were readily apparent. Corporate entertaining and politicking one's way to the top were famously the primary Exec' activities in the Tech Divisions, where actual results were secondary. The argot and culture between NDC's various Divisions differed, the outcome of multiple Corporate acquisitions and failed blending; Commercial Imperialism.

Land Division was an unlikely place in which to find career-conscious Executives, as the work was too hands-on and too immediate, plus, there was always the possibility of having to visit a Conflict Zone, which these Personnel sought to avoid at all costs as it was not sufficiently career-enhancing. None of them were aboard the C130 which flew me into Kuwait over the Minefields and burning Oilfields, and none of them were to be seen later in the Baghdad Green Zone or in Tamil-beset Sri Lanka either. I enjoyed the hands-on Customer and Agency contact which was essential for success and I liked winning. There was a very small handful of truly successful international Defence Sales Executives and I knew most of them.

Directors, Shareholders and Employees all need big overseas Defence Orders to bring in high-margin revenue, and the UK Government wants them desperately to sustain jobs it claims, especially in Labour's morally vacant new world of '*Defence Engine Britain*', an almost exact throwback to the Wilson Government ethos of the Sixties. But few individuals are either prepared or equipped to do what it takes to win this business. Watching from the touchline has become a national affliction, but everyone wants to be pictured holding the Cup. Whatever high-sheen Corporate gloss and posturing NDC might apply for the benefit of the outside World and their Shareholders, the dirty business of winning Defence Orders remained their bread and butter.

NDC's marketing theorists remained handicapped by a feckless belief that overseas Customers would continue to buy British Defence equipment automatically, simply because it was British, and because we told them it was the best. This breathtakingly arrogant posture was of course plain wrong, as successive Gulf Wars, Iraq and Afghanistan have demonstrated. It was as if NDC's Marketeers had missed everything: the fast-rising tide of foreign competition, international consolidation of the Industry, key geo-political sea-changes in marketplaces, and new attitudes to Arms Cash; all these had seemingly passed them by. So cold commercial waters lapped around NDC's toes, whilst Canute-like, they faced a rising tide with immobility and inflexibility.

Way-up high in the Corporate and political stratosphere, someone must have realised this, and recognised that the continuation of Agency and Arms Cash mechanisms remained the key to sustaining UK Defence exports; the other key was to encourage foreign owners to take over British Defence Companies, which they duly did by the cart load. Successive UK Governments fed foreign-owned Companies with UK Defence Contracts, Taxpayer Cash, IPR and incentives, so national Defence Technology and manufacturing capabilities diminished sharply, and innumerable Defence jobs have been exported. It's has been staggering to watch it all poured away to the detriment of UK Forces and British taxpayers.

Foreign shareholders also benefit from the massive British Political gifts of Defence equipment and weapons to Ukraine, where, at time of writing £7.8 billion of UK taxpayers hard-earned cash was promised by way of military assistance. No Corporate Sales efforts or expenses required there, as witless Politicians simply gifted taxpayers' cash to the overseas Shareholders of the manufacturing companies concerned. Stormin' Starmer's cringeworthy '5000 *missiles for Ukraine*' announcement was a great example. Primary beneficiary of this £1.2bn freebie? The French Government, which holds the majority stake in Thales, the Belfast-based manufacturer of the missiles being gifted. I feel for the US Taxpayers who have paid out billions of dollars just to keep a foreign war going, but at least most of their Corporate beneficiaries are US-owned. It's not so in UK, most of the cash leaves the country.

The market reality of the non-competitiveness of so many UK Defence legacy products at that time, served only to emphasise the importance of key channels and individuals such as Hector, if any business at all was to be won. There can be no other explanation as to why successive UK Governments consciously avoided crackdowns on major Arms Cash-paying Defence Companies. Governments were happy enough to jail a few Company Minnows who got caught in the legislative net from time to time, pour encourager les autres, but the major UK Plcs needed their Hectors and they swam free.

The dirty-washing incident with a previous CEO ensured that NDC trod very warily in-country and had thus missed out on major deals, so the new Hector channel could re-ignite an NDC dream and put the Company back on the regional map. Business on a truly strategic scale was possible. It was impossible to get as much as a cigarette paper between some senior NDC Directors and certain Government Departments and Ministers, the dividing line between them wasn't just blurred, it simply didn't exist. Sadly, though the blind were leading the blind, so aspirations and operational realities remained 'out of sync'.

I might have upstaged NDC's Corporate Marketeers by getting close to Hector where they had failed, but Hector would have to be controlled at the highest level in NDC, such was his status and strategic value. I would remain a key working-level contact in the field for the time being, until re-tasked, but the working relationship we established was to prove critical for all of us quite soon. I fully expected to be 'dropped like a hot brick' by NDC when it suited them, as Senior NDC Directors wanted a slice of this action for themselves. In closed session meetings with no more than three or four people present, they smiled and congratulated me on establishing the channel, but I was under no illusion that in the corporate politics of it all I was now their enemy, for I had succeeded where they had failed. Establishing Hector as a key channel also denied Senior Executives the opportunity to sign-up their own favoured Agent candidates in-country, from whom they could have derived lucrative kickbacks.

Shortly after the London de-briefing session NDC announced (internally) a substantial new Defence contract, the first and smallest phase of which was worth over fifty million pounds, and which represented a very nice pay day for Hector who stood to make over five million in total from this phase. This value of the deal was not on the mega-scale of say, an aircraft deal, but values are relative and for the Division concerned it was huge. It was incremental high-margin business and it was won and 'brought in' despite the NDC Commercial Department's efforts to f*** it up. They were almost a bigger challenge than anything I had to manage in-country, except avoiding being chopped-up of course.

This phase was the only first contract in a series which would total over three to four hundred million sterling, and which also paved the way for a five-year rolling-programme which Hector and I were configuring. It was significant and it was strategic for NDC. My Boss rewarded me for my efforts with a bottle of cheap sparkling wine which I dropped into his waste bin on my way out of his Office, suggesting to him that he shoved it up his arse. He'd never been a Sales Executive, in fact his appointment as Sales Director was his first ever Sales appointment. This is not uncommon today. Oh! to be in an American Company at a time like that.

Before aligning the second phase of the deal, by then a revised multi-phase package totalling over £80 million, I took a much-needed break as I was tired from travelling around the Middle East, and I'd been working on deals for other countries at the same time. The constant clandestine manoeuvring in-country was very wearing, especially when executed with such intensity. Whenever I was back in the UK Office there was a constant need to watch my back as I was aware of the plotting and scheming which was going on with other NDC Executives seeking to emplace competing Agents in the same Territory. It's taboo in a Defence Company to brief people on Agent's identities, as these are tightly kept secrets, so I was also very uncomfortable because 'fat-boy' East seemed to have become suddenly very free with some of my information, especially amongst his Corporate Glamour-boys, who had their own agenda. In any case, it was going to be great to escape from it all and I headed off to Ravello for my break.

My return from Italy was delayed by a cancelled flight, so I took a pleasant extra day in Naples, but when I finally got back to the Office the shit had well and truly hit the fan and most of it seemed to be heading my way. Fat-boy East led the charge. It was highly unusual for old 'Two-lunches' to be found at Land Division's scruffy provincial Midlands site as this was way outside his London Lunching-Zone. NDC's Corporate Marketeers seldom strayed outside the City, where their intimate knowledge of London's finest Restaurants was honed with a frequency which would shame any restaurant critic. These Marketing Giants needed only the merest hint of an excuse to lunch or dine out at Corporate expense with any passing overseas Delegations, London-based Defence Attaches, Industrialists, Civil Servants or other convenient persons, however obscure they may be. They grazed at will 'on the Company' throughout the City and venturing outside the M25 boundary to a provincial Midlands Divisional Office was an undesirable, almost unbearable burden for any of them; a punishment posting almost.

As soon as I arrived on-site, late as usual for a Delivery forecast meeting, I was summoned to East's borrowed office where he sat alone, corpulent, sweaty and twitching behind a desk. There were no greetings and he didn't look up when I entered the room, so I sat down in the vacant chair opposite him, fully sensing that 'something was up'. How depressingly right I was.

"There's been a bit of a problem…" East quivered eventually, still avoiding my eyes and staring down at the blank pages of his notebook. There was a ghastly pause and his usual florid complexion reddened and deepened further. I thought I'd have to call a Medic; he was surely about to burst. The words trickled out slowly, East could barely bring himself to say it aloud.

"Hector's Commission advance is missing… three million pounds of it…" East's unhappy fear-filled eyes finally met mine. Hector's Arms Cash, partially paid up-front as agreed, had indeed gone missing and according to Fat-boy East Hector was not a happy man. Well, there's a f***ing shock! How happy did East expect Hector to be after losing his three million? This is a Defence Corporation's worst nightmare; an illicit Arms Cash commission payable to an overseas Royal Family member goes missing. Exactly what can be done about that? East looked away unable to hold eye contact even momentarily, something else was coming, I sensed it, so I jumped in first.

"Then we need to see him quickly and sort it out..." I responded. Silence from East. He dribbled out a few more words, the big one was coming.

"...it's being said that it's, er, an inside job... an NDC inside job..." The penny dropped with a resounding clang. East was all but accusing me of taking Hector's millions. I challenged him to say it again, but he declined.

"Wait here, I'll get a Lawyer and you can tell him..." I said furiously. But East was very, very anxious to keep the lid on things.

"There's no need to do that..." East urged, and we started to discuss it properly and out it all came. There was no question that the money really had gone. The cash left NDC's account but Hector's nominated Bank Account hadn't received it. East revealed that just prior to the Bank transfer, NDC received instructions to amend the details of the receiving Bank and sent the cash to this newly nominated Account. The original Account in Hector's Agreement was apparently a Swiss bank, although I'd never seen the details, as his Agreement, like all NDC others, had been signed in Switzerland with Agency Director Harry Kensington. The new Account described by East was apparently in an American Bank. An American Bank? They must have been mad; big red flag. No one sends Arms Cash funds from a UK Defence plc., or its ghost, to the USA, because of the money-laundering regulations. Idiotically, NDC tried it and the nominated US Bank predictably rejected the transfer.

Favourite banking locations amongst Arms Cash Companies varied in those days but included the British Virgin Islands, home of the infamous Red Diamond account, Liechtenstein, Lebanon and of course Switzerland, although nowadays this latter location is less reliable that it used to be, owing to changes in Banking regulations there and the newly acquired piety of a new generation of compliant tell-tale Bankers. The USA is not on any Arms Cash list of preferred banking locations, and any transfer of funds from a Bank Account traceable to a leading UK Defence Corporation would stand out like dog's balls. I wondered who in NDC had verified the change of details with Hector, and I asked East exactly this question. Sheepishly he conceded that no, NDC hadn't confirmed the details with Hector in person, in fact NDC Executives had not verified the new instruction at all, they'd simply accepted and acted on a 'documentary instruction' whatever that was, without validating it. Schoolboy mistake. Could a major UK-based Defence Plc's really be this stupid you might ask? Well yes, they could, yes, they were.

Instructions involving any changes to Agency Agreements, such as Account details, are only ever validated face-to-face with the Principal, and the gibbering East would have known better than anyone in England how and why this was so. It was a core protocol especially for an Agreement of such importance and sensitivity, and it should have meant a trip to Switzerland for Harry Kensington. For unexplained reasons the NDC apparatchiks accepted the change notification without a murmur, but worse, if that was indeed possible, was to come.

Flustered by the US banking rebuff, NDC bureaucrats and financial wizards, ever anxious to please their bosses by getting the job done, requested an alternative account number from the same source (which turned out to be a Fax number). It was as if the failed transfer to the US had been their fault and they wanted to put it right as quickly as possible. They didn't refer the issue either to Harry Kensington or Fatty East, they just wanted to get the funds moving as quickly as possible just as they had been told to. Fearful of bollockings for their botched attempted transfer, the Finance team accepted a further 'documentary instruction' (another Fax) from the same unvalidated source, and this time the Transfer was mandated to an Australian Bank.

The second Arms Cash transfer was completed cleanly, and the NDC Arms Cash funds moved across the world to the new Australian receiving bank, from where they were whisked away instantly to somewhere else. The Arms Cash passed straight through the Australian Account and kept moving and apparently landed in India. It was breath-taking, not only had NDC twice responded to unvalidated changes of hyper-sensitive and critical Agency bank account details involving Banks on two Continents, but three million pounds of NDC's Shareholder's cash simply disappeared into the ether without ever reaching the nominated Arms Cash Agent, a Middle Eastern Royal Family Sheikh. A Royal Flush then in every respect.

Fatty East became even more sweaty and twitchy, and so he should have been. I pressed him for more information about the Account change, but it was a very raw nerve. Executive faces weren't simply red with embarrassment, white heat glowed and flowed from Finance and Commercial Directors and a pyroclastic flow of mutual recrimination and finger-pointing followed. There was another of those ghastly long silent pauses between Fatty East and me. He was in a state of shock, and frankly if he wasn't then he should have been, and not just because he'd had to leave London and to visit the Midlands. I knew the exact technical term with which to summarise this critical situation.

"It's a complete cluster-f***...!" I said, lighting up a Marlboro in his 'No smoking' borrowed Office. What to do next seemed obvious to me.

"...we have to get out there soonest and resolve this with Hector in person and we have to do it now...."

Two-Lunches was approaching his melt-down point. He was clearly out of his depth in this crisis and I thought he would suffer a seizure. He front-loaded his excuses for not going personally and said it would be 'inappropriate' for a Main Board Director for some unexplained reason. Apparently, he also had some other, undefined, urgent stuff to do. Yes, of course he did. More urgent than recovering NDC's three million I wondered? The Finance Director also declined to go as he was needed at home he said. He was terrified at the prospect of having to travel outside England, especially to mop-up this mess; his mess. In his worldview, the Middle East was somewhere where only the Sales and Marketing teams needed to go. NDC Senior Executives were in a state of Corporate paralysis.

NDC loved the money it made in the Middle East; NDC needed the money it made from the Middle East. The Corporation charged up to three hundred per cent more for its Systems there, than to domestic customers, and Directors thrived on their fat bonuses which such orders funded. But none of the Board ever went to the Region except to showboat during Exhibitions or languish idly in Dubai luxury Hotels with Asian prostitutes. It was altogether too uncertain and too dangerous a place for them. I once won a contract worth nearly sixty million pounds in the region and could not persuade a single Company Director to come out and sign the showcase contract with the Defence Minister. Now, as usual, I seemed to be the only one willing to make the trip and that 'stitched-up' feeling set in. I had to prompt East again because he was going comatose and he needed a sharp jab. I just wanted to punch him really.

"Send whichever Minder you want to go with me, but let's get on with it, let's meet Hector and sort this out..." I said, adding "...the longer we leave it the worse it'll get..."

East didn't pause to consider my offer longer than a nano-second, it was his dream. A Corporate Minder was duly assigned, flights were booked, and to my great surprise I was going to have NDC's A-team for company in the shape of Harry Kensington himself. The perfectly proper, perfectly pukka Kensington was Sir CEO's trusted right-hand man in all senior Agency matters, and this shadowy Gatekeeper of NDC's biggest and dirtiest Corporate Agency secrets was coming with me. No one outside NDC had ever heard of Kensington, and internally his role was known to only a handful of senior Directors with good reason.

It was Kensington who flew to Switzerland, the Virgin Islands or elsewhere, to meet NDC's most sensitively appointed Arms Cash Agents and manage their Agreements. These Agents were emphatically not the lower-level 'Informers' I often met, but rather heavyweight Influencers, many of whom were foreign Royal Family members or Government Ministers in their own countries; people just like Hector in fact. It was Kensington who held their hands as they counted out their NDC Arms Cash Commission millions into their discreet offshore Bank Accounts, and now he was coming with me. To hold my hand or what? I wondered.

Proper and undoubtedly pukka he may have been, but Kensington had a thoroughly improper Secretary, Katrina, who managed not one but two fat little black books of contacts, stuff never to be trusted to digital devices, which included an index of top London Madams. Katrina ensured that visiting Defence delegations and tame Agents were entertained appropriately, i.e. inappropriately, whenever they came to Britain. NDC's Hooker bill was substantial with every conceivable need being catered for, although it was going to take so much more than a few high-end whip wielding Dommes to get us out of this mess, however adept they might be with their 'specialties'.

The surprise involvement of Kensington reflected Hector's strategic importance to NDC, for Hector represented the path to the golden uplands of strategic business the Company wanted and needed so desperately. The three hundred million pounds worth of potential business I was now projecting, would be financial chickenfeed compared to what the rest of NDC could get out of this rich and high-spending State with the right Influencer on the Arms Cash payroll. UK political moves in the territory were ineffective as usual; politicians couldn't deliver Newspapers for UK, let alone Business, and UK was being comprehensively outplayed by French and US Companies and Governments. NDC needed Hector.

But there was another reason, a secret corporate political reason for Kensington's surprise involvement, although at the time I didn't know it; another murky little NDC surprise was hidden in the geo-corporate undergrowth waiting to jump out. For now, all I knew was that I had a new and unwanted Main Board travelling companion, which made an unwelcome change from my usual modus operandi of travelling and working alone. Adjustments would have to be made; I would have to behave myself.

The Board needed to determine a strategy for resolving the matter of the missing millions, so two high-level emergency meetings were held in London before Kensington and I left for the Middle East. Ultra-sensitive issues had to be managed. I was glad that accounting for three million missing pounds from NDC's Corporate Accounts was outside my remit, and it must have created an interesting accountancy challenge for a Plc with formal Accounting, Audit and Compliance obligations. Just how did they manage to account for these clandestine Agency payments and how did they square them with Auditors? What would they have to say about this f***-up?

"We paid out three million in illicit Commission, but someone stole it..."

Oops! Love to be a fly on the wall of that conversation. I began to get a feeling that this wasn't NDC's first missing money Rodeo.

The pre-departure meeting group was necessarily kept very tight and no Minutes were taken; this was strictly unattributable word of mouth stuff. Attendees included Fatty East, NDC's Marketing Director, plus Harry Kensington, Sir CEO's right-hand Agency man, plus NDC's Main Board Finance Director. But at the last moment, this group was joined by the super-slimy Jolyon Tennyson, NDC's Senior Middle East Marketeer. It was upon Tennyson's delicate corporate toes that I'd landed with such an audible crunch when I'd won big business in a Country where his own team had failed so badly, claiming there was no available Defence budget. I didn't trust him an inch. I'd first met Tennyson in his London NDC Office two years earlier when he'd tried to recruit me into his team, but I'd declined. He tried again later.

This select little group discussed the missing Arms Cash situation and considered remedies, and it was Harry Kensington's job to brief Sir CEO afterwards; a Taxi waited outside ready to whisk Kensington away for this purpose. A simple plan was hatched. Kensington and I would fly to Bahrain, neutral ground, and meet Hector who would be assured that NDC would pay the three million Advance again, but needed additional business, as the existing Contract couldn't bear this level of extra cost. Hector would have to guarantee NDC extra business to make up the difference. Between us, Hector, Kensington, and me, we would try and work out what had gone wrong whilst NDC's financial Wizards would track the missing money and try to recover it.

Greedy NDC saw this as an opportunity to sell-in some additional high-value Systems where they had failed before. I would set up the Bahrain meeting with Hector's Egyptian Lawyer Essam, whom I'd met during my early meetings with Hector, and with whom I sustained back-channel communications. Essam and I got on well together, which was fortunate, as it was critical to involve only people we trusted. Shortly after the meeting Sir CEO approved this 'plan' and Kensington and I made ready to fly out. But just before we left, there was an odd development.

Slimy Tennyson, who coincidentally had been 'in-country' at the time of Hector's cocked-up cash transfer, and who I later discovered had first pointed the finger of suspicion at me for the missing cash, introduced another Arab businessman into the fray. This conflicted with the need to keep discussions tight. The new individual was an Arab Merchant called Saeed and he was known to be no angel, in fact, he had 'form'. I didn't know how or even if Saeed fitted into NDC's great scheme of things, maybe he was an existing NDC Agent or similar, I wouldn't necessarily know. For all I knew, he was someone's 'Runner'. I had to accept his presence, but I distrusted him from the word go.

I learned discreetly that Saeed had once claimed to have been the victim of a failed kidnap & ransom attempt years earlier, although it remained unclear, even after the ensuing Police investigations, as to whether or not he had been a genuine victim or if he had actually stage-managed the whole thing himself to extort a ransom. The Defence Industry is awash with such cuddly characters.

Saeed and Jolyon Tennyson seemed to be very close, much too close frankly; maybe they were in a personal relationship, who knows. Kensington told me later that Tennyson had once tried to have Saeed appointed as an Agent in Hector's home country despite any evidence that he could deliver NDC business. Saeed's apparently high status in the local Merchant Community didn't ensure or guarantee influence or capability in the Defence sector in-country; these were very different domains in many ways. Probably for these reasons, and there may have been others, Tennyson's earlier efforts to achieve Saeed's appointment as an NDC Agent had been resisted by NDC, which at this level meant that Sir CEO himself had vetoed Tennyson's proposal. A high-level Corporate power struggle was in progress. Something else to be very careful about.

Saeed was introduced to the meeting by Tennyson and Harry Kensington seemed to be as bemused as me at his unforeseen introduction. It turned out later that Saeed had made a secret proposition regarding this Arms Cash crisis directly to Sir CEO. Saeed's concept of operations was to 'negotiate' personally with Hector to reduce the Arms Cash commission which NDC had to repay. Saeed claimed that this would be such a valuable service as to merit a fat Arms Cash fee for himself, expressed as a percentage of a downwardly revised commission figure, which he would negotiate with Hector. Saeed proposed to work thereafter as NDC's Agent and manage all future opportunities covered by an NDC Agency Agreement. It stank up to high heaven and back again twice. Slimy Tennyson and Saeed were joined at the hip and wallet.

Kensington tipped me off that Sir CEO distrusted both of them, and that and it was obvious that Tennyson had briefed Saeed on the missing Arms Cash. How else could Saeed have known enough to make his direct approach apparently out of the blue to Sir CEO? I was certain that Saeed didn't know Hector personally and Hector was far too canny an operator to let slip a story like this one about missing Arms Cash, especially to a Merchant of non-Royal status. It was supposed to be a time for extra discretion, not less. Harry Kensington, and therefore probably Sir CEO too, speculated that Tennyson and Saeed were in cahoots in planning to share the commission which Saeed proposed to extract from NDC. It was a complication which Kensington and I could have done without as we set off for Bahrain, and Saeed's proposal was put firmly on ice at the direction of Sir CEO, who clearly wanted nothing to do with it, for reasons I was not party to.

It was an uneventful and sparsely occupied flight to Bahrain. Kensington, a quiet man by nature, buried himself deep in some dreary historical novel whilst I read all the newspapers, flipped through most of the magazines and watched two films, trying unsuccessfully not to think continually about what would happen when we met Hector. I'd no idea how he might react, so mentally I gamed various outcomes. All too soon, it seemed we landed, and with the Immigration channel smoothly and swiftly negotiated, we were collected by a hotel driver for the shortish drive to the hotel. We checked-in quickly and went up to our rooms to freshen up, but I'd spent barely a few minutes unpacking my computer when the house 'phone rang. I thought it would be Kensington, but I was wrong. An unknown but unmistakably Arab voice speaking in calm and measured tones advised me that...

"We know why you are here and what you have come to do. If you continue it will be terminal for you personally and for your Company..."

I asked the caller to repeat the message, which he did, and I fired off questions, but he hung up. I dashed along the corridor to bang on Kensington's door. There was a delay and then I heard the lock being turned and the door opened. Harry Kensington looked like he'd seen a ghost.

"Did you just get a weird 'phone call?" I asked him.

He had. It was identical to mine; our task was well and truly compromised. Kensington was not a hardened Business traveller in the sense that I understand 'hardened business travelling'. His Club Class forays into Europe didn't count as far as I was concerned, and neither did his First-Class long-haul sojourns to the Virgin Islands, etc. It was one thing to meet up with already wealthy NDC Agents in five-star Hotels in distant overseas locations, or in Private Banks or in expensive Lawyer's Offices, where Kensington might help them eat their caviar and count out their Arms Cash, but it was entirely another to have your life threatened in a Middle Eastern State, barely minutes after checking in to a hotel.

The blood really had drained from his face and he slumped in a chair in a state of shock, probably asking himself how on earth and why, he'd come to be here at all. It was a 'Dry' room in a 'Dry State', so I couldn't pour him a consoling drink. I used Kensington's room 'phone to make a prearranged call to Essam, Hector's Egyptian Lawyer, to confirm that we were on track for the meeting, which was to take place in the Holiday Inn a little way from our own Hotel. Essam answered the phone and I paused momentarily to glance down at the shell-shocked Kensington. I decided.

"See you there Essam" I said. I wasn't going to have a debate about it. Kensington and I discussed these latest threats over Turkish coffee in the hotel's Coffee Shop, and he agreed that as the meeting was both imminent and geographically close to our hotel we should go ahead, but afterwards we'd lose no time in getting straight back to the airport, even if we had hours to wait for a flight. It would be more difficult for anyone to have a go at us if we were airside, with Airport Security and Police in proximity; these folks might represent at least a token deterrent. Whilst we were static and still in our hotel, we would be the easiest of targets. My usual mantra applied; 'having limited options does not mean do nothing.'

We checked out of the hotel and took a cab as if we were going into Manama, i.e. the wrong way, then changed cabs and turned back instead to the Holiday Inn which was close to the hotel from which we'd started out. We went to the appointed floor and room but not directly, taking firstly the Lift, from which we exited two floors early, then up the fire escape stairs. We walked past the room initially, pausing only to listen, then returned to it and I tapped on the door; three brisk double knocks as agreed. It was opened instantly by Essam who ushered us in, glancing left and right along the corridor as he did so.

Hector was seated on a sofa backed against an interior wall and the curtains were drawn. Essam invited us to sit down; coffee was already on the table and TV sound was on to fuzz the background. Unsurprisingly there was a tense and prickly atmosphere, it felt awkward. I introduced Hector to Harry Kensington, who would normally have met Hector for the first time in some safe and luxurious offshore haven to sign up his Agency Arms Cash Agreement. But the tempo and irregularity of events had disturbed the usual covert pattern and we were all on edge.

This meeting was a stark contrast to my first meeting with Hector in his discreetly located Villa, where together with Abdullah we demolished his bottle of Black Label. On that occasion he'd enjoyed and 'played' the theatricality of our encounter, but now he was extraordinarily tense; he just wanted from NDC what was his, what had been agreed, the three million Arms Cash advance. He'd kept his half of the bargain and delivered the major Contracts and now NDC had to honour their half of the deal and make the Arms Cash payment. As far as Hector was concerned it was black and white, it was all NDC's fault. Hard to disagree with that.

Both NDC and Hector needed to get something out of this meeting, but the ever-arrogant NDC Executives back home had not gamed Hector's position very well; they were intent simply on saving their own bacon. Hector understandably wanted his money and ran considerable personal risks in his own world by aiding NDC. Exposing a senior figure like Hector in his own country would be political dynamite and could easily result in lethal consequences. You must tread softly.

If NDC was exposed in the UK Press it would also be explosive commercially and politically with a lot of blood on the carpet. NDC harboured strategic aspirations in Hector's homeland and desperately needed to fund the repayment of the stolen commission, if indeed repayment was to be the outcome. I also needed something personally from the meeting, as I wanted Kensington to hear for himself and directly from Hector, that he didn't hold me responsible for the loss of his Commission, his Arms Cash. Fatty East had talked about an 'Inside job' when he first told me about the lost Arms Cash, and the egregious Tennyson had probably fanned the flames of that little story. I guessed it was also part of Harry Kensington's Brief from Sir CEO to determine if someone inside NDC was involved in this debacle.

All things considered the meeting got off to a bad start, although not as bad as a meeting in UAE once, when an Agent I had to 'release' pulled out a handgun as I was talking to him. Essam spoke for Hector who remained silent, he simply and abruptly demanded payment of what was due, what had been stolen, and we talked around the subject for a while but it was obvious that in this frosted stand-off atmosphere we would get nowhere. There were long silent pauses and hard looks were exchanged. I decided to take a chance on starting a real dialogue, so I addressed Hector personally and directly.

"Look, we only did any business in the first place because you and I trusted each other, so let's get back to that. I'll tell you everything we know about this, let's try and work out between us what's happened…".

Harry Kensington looked a bit worried about this, but it proved to be the game changer. Hector broke his silence, the frost thawed, and we compared notes about what each of us knew. Both sides knew the theft of the Arms Cash had been an inside job, and it looked to Hector like an NDC Inside job, but as we pieced it all together it became increasingly obvious that there had been activity on both sides of the fence. I didn't mention to Hector our telephoned threat in the hotel across the road before the meeting.

After we'd gone over it all a dozen times or more and discussed every possibility, we agreed that NDC would pay again, but that there had to be some additional business to soak up this repayment. This was exactly how we'd envisaged our position before we left UK and at length Hector agreed, subject to the repayment being prompt and with a Commission increment next time by way of compensation. He also reassured Kensington that the theft of his Arms Cash couldn't have been my doing.

It was agreed that Essam and I would meet up in Cairo to progress things and to determine a new schedule, since the repayment of the whole Arms Cash amount could not be achieved in a single lump. We'd also work out the details of the new contracts through which the repayment might be recovered by NDC. Essam and I would also coordinate the planning for some new-new business, not all of which was in Hector's home country. Kensington and Hector would meet separately in Switzerland to formalise the Agency Agreement, together with protected new Bank and payment details. Both Parties remained understandably edgy after the initial frostiness, but I felt we would move forward again. After agreeing everything with Hector and Essam, Kensington and I left the Holiday Inn. It was probably the best outcome we could have realised in the circumstances.

The threatening 'phone calls had persuaded Kensington and me to head off in a Taxi directly to the airport, for the very long wait for a flight, seven hours in fact, but Airport Staff allowed us through to the empty Business Class Lounge which was a big relief. We drank a lot of coffee, discussed events over and over, and the same thoughts regarding the 'threat of the day' occurred silently to both of us. Kensington was the first to vocalise the subject.

"Have you thought about how our Hotel calls originated?" he asked me. Oh! yes, I certainly had. We discussed it, logically considering each event and all the players in this grubby drama, until we reached a natural pause. Kensington had a good brain for this sort of thing and would have made a better than average Spook (in fact he liaised with them on behalf of NDC, I learned later) and I guessed he probably dashed off 'The Times'

Crossword fairly swiftly over breakfast without breaking any mental sweat.

"I've got a name..." he said, "...me too" I replied. We agreed to write down the names then exchange them across the coffee table; a bit childish, but it seemed like a good idea at the time. Well, we were very tired. We read each other's selection. Same name; Tennyson.

The return flight to Heathrow was as smooth as empty and as uneventful as the outbound journey and Kensington resumed his reading, whilst I trawled for any magazines I hadn't read on the outward flight and flicked through the film choices I'd rejected previously. I also worked my way through a bottle of red wine with dinner, whilst Kensington slept, snoring loudly. On arrival at Heathrow T5 in the early hours of the morning we positively floated through Immigration and exited the Terminal together.

"Driving home?" I asked Kensington. But almost as I spoke, a super-glossy NDC Mercedes appeared in the pick-up zone.

"Got to brief Sir CEO..." said Kensington as he climbed into the car, which sped off with him into the misty early morning, towards London.

I was supposed to have the kids again that weekend, but I just couldn't do it, a second death threat in the Middle East and a treacherous NDC Executive were more than enough to occupy a very tired me for the whole weekend. My Ravello break already seemed like an eternity ago; I needed another one. Never before or since, have I more greatly wished for a regular, boring nine-till-five job. There must be better ways to earn a living I speculated. H'mm, maybe there were, but what about the 'Rush'? Wouldn't get that working for Marks and Spencer.

Ten days and lots of arranging passed before I was back in the Middle East, this time to meet up with Essam in Cairo, and this was an altogether more pleasant trip than the difficult meeting in Bahrain. I'd stayed in the Heliopolis Sheraton during previous Cairo trips, but I stopped using it after the fire there which killed sixteen people. I stayed instead in the Nile Hilton, ironically, where I'd once met up with the Former Ambassador Extraordinary and Plenipotentiary. I had a wonderful night-time view over the ancient River of Life, and Essam was good company. We dined in the top-floor Restaurant where we also watched the Show in which the Egyptian drumming was so deafening that it was hard to hear the chat.

We were getting business issues firmly back under control and the next phase was already in progress, but Essam shocked me. He revealed that shortly before the meeting in Bahrain, he and Hector too had received threatening phone calls much along the lines of ours. Like Kensington and me, Hector and Essam too compared notes to work out the likely source of their own threats, and they too had settled on Tennyson, along with his in-country buddy, the wannabe Agent Saeed, as the likely perpetrators.

Hector's significant personal influence in National Intelligence made it relatively easy for him to collect corroborating electronic data on Saeed, as Hector's Western-trained Intelligence Technicians were not burdened with the same restrictions as their UK counterparts regarding intercepts etc. They also identified Hector's Indian Manager Bandar as the probable administrator of the account details switch, which could have been either coincidental, or an additional event to the Tennyson/Saeed cash-grab plot and Bahrain threat. Bandar had subsequently 'done a runner' and disappeared completely. Quelle Surprise! In fact, he'd disappeared to India. It wasn't clear if Bandar had been working with Tennyson and Saeed from the start, or if he had just spotted the main chance for himself and acted alone to rob both his employer and his confederates. Murky, even by Arms Cash standards. I confess that I knew a bit more about Bandar than I'd let on. Remember, in this business, always keep something in your back pocket.

There were consequences of course. There are always consequences. Hector 'leaned' on Saeed and put him back in his place, just as he had done with the Former Ambassador who'd threatened me. When I heard this news, I recalled the cowering disposition of the Former Ambassador during our City Office meeting, in which he'd apologised profusely and grovelled. I wondered if Hector had exerted the same charm and influence over Saeed, who'd crossed a line and deserved whatever Hector dished out to him. No sympathy from me. Tennyson was removed quietly from his NDC Middle Eastern role, but couldn't be sacked outright, because as one of NDC's senior Middle Eastern Marketeers, he simply knew too much about NDC's dirty little secrets in the Region. It's simply a fact of life in an international Defence Company, that knowing enough can make you fire-proof, knowing too much makes you dangerous, but knowing nothing makes you useless and vulnerable. You have to find the right balance. 'MAD' at work again, Mutually Assured Destruction, with each party, in this case Tennyson and NDC, possessed of enough knowledge about each other to assure destruction or critical damage through exposure; it's a form of Defence Industry S & M. Tennyson eventually resigned a year or so later and was doubtless encouraged to go by NDC with a soft goodbye. He virtually disappeared from the scene, save for a Consultancy phase, which is what we all say when we're unemployed. I doubt that he missed the salary very much, as he was doing well enough without it. But the world of Arms Cash is acted out in the Hall of Mirrors, so here come the reflective twists.

NDC eventually tracked down the missing Arms Cash which had fetched up in an international Bank in India, and they sent out a team to try and recover the cash. Imagine the scale of that challenge though; you must explain to this Indian Bank, firstly, that your major UK-based Defence Plc has made an illicit Arms Cash commission payment of three million pounds to the wrong bank account, well, twice in fact. Then you must ask an innocent third-party bank, who didn't receive the money directly from you in the first place, 'Can we have our money back, please?' What are you going to do if the bank says no, sue them? I don't think so. Then, how do you patch over the £3m hole in your Corporate Accounts and what story do you, or your Corporate Accountants, tell your Shareholders and Auditors?

Bandar disappeared, never to be seen again and I neither saw nor heard of the Ambassador either, despite being a frequent visitor to the same territory on many subsequent Defence and other business visits. The story I held back was that I'd met Bandar in London when he'd unexpectedly come to England as we were positioning contracts for NDC with Hector. Bandar had claimed then to be travelling on business on behalf of Hector, which seemed a bit far-fetched. Why would Hector, who enjoyed significant and sophisticated international business interests and contacts for his non-Defence businesses, send a low-level 'Gofer' like Bandar to conduct any business on his behalf? It didn't fit.

I met up with Bandar at his request in a West London Hotel a couple of times during this trip, where he appeared to be entertaining a series of international Arms Suppliers, including some Russians who I encountered in his hotel corridor one night. He was exceeding Hector's brief, whatever that may have been, in a big, big way during this trip, becoming in his own fantasy world the 'Principal' instead of just being the Messenger. It's a common mistake amongst Runners. Ever-optimistic and naive UK-based Defence Companies had also made the mistake of treating Bandar as a Principal and had pandered to him accordingly. He loved it, he lapped it up, and played it well. Some UK Companies actually signed up with this Gofer instead of his Principal. More fool them. It later went very bad for at least one of these Companies.

Apparently eager to impress me with his worldliness and international Defence business savvy, Bandar let slip in a very drunken moment (he could not cope with Whisky, but could not resist it either) that he made secret recordings of his various Defence business meetings using a small voice-activated tape recorder (he was not alone in this habit). His purpose in making these recordings was, he claimed, to save them for 'personal Insurance' purposes. He boasted that he'd also used the tapes to blackmail Companies or individual Executives and 'Brokers' who had agreed to take kickbacks. Was this dangerous madness or just bravado? Maybe it really was true, and it prompted a question in my head, had he recorded me?

Bandar was frequently drunk during his London trip, and when a colleague and I visited him one evening as arranged, he failed to answer his room door. We accessed his room and found him deep-sleeping on the bed fully clothed in an apparent drunken stupor. We also found a powdery heap of what seemed to be Cocaine on the bedside table. The papers and notes spread out across the table and strewn all over the floor bore testament to the fact that Bandar really was substantially out of his depth. He was, or so it seemed, involved or trying to get involved in a few Arms purchase deals which bore no relation to those I was working on with Hector. I doubted if Bandar would have been able to deliver any serious business, either within the Defence domain or otherwise, and he was raising foreign Defence Company hopes unrealistically. Bandar was leading some very serious Arms Cash Players right up the garden path; not a survivable strategy.

Aside from the various papers and Cocaine, we also found used mini-recording tapes, and most intriguingly, on a crumpled munitions computer print-out found with the tapes, was a familiar handwritten scrawl, it was that of a certain former Ambassador. Bandar, it seemed, had also been one of the Ambassador's secret 'Runners' but had gone rogue on him as well as doubling on Hector. Bad enough, but this wasn't the only surprise. In a very neat twist, one of Bandar's recorded tapes revealed a secret meeting between Bandar and none other than the slimy Jolyon Tennyson, and it was an interesting taped discussion about kickbacks. I'd once been courted by Tennyson and his Middle Eastern team because they wanted me to join them, and that bizarre Interview is worth recounting, because this sort of incident could occur only in the paranoid and f*****-up Defence Industry. The prequel to the 'Interview' also provides a glimpse of the well-meaning-but-stupid-at-work on Defence Sales in the Middle East. Only the British could, or would ever do it like this.

NDC had arranged a Drone Interceptor demonstration, a very high-profile event in the Middle Eastern State concerned, and I was to head-up an accompanying 'mini-exhibition' showcasing other NDC systems as a side-salad attraction to support the event. Demos like this are hellishly expensive to execute, involving numerous Staff, high-technology hardware, travel and transportation costs. The Interceptors featured some fizzy new technology which didn't come cheap.

The site allocated for the Demo was a Desert Range far from urban hotel-land and creature comforts, and NDC's corporate marketing glamour team was on duty to host the invited senior military audience, plus a few neighbouring-country Sheikhs. It occurred to me, and no doubt to the VIP audience too, that running this Demo' on a Desert Range at close to midday in the Summer heat was perhaps not the most convivial way of conducting the event.

NDC's regional office team arranged for tiered seating to be erected for the Guests, with VIPs in comfortable armchairs at the front. There was no canopy or shelter from the hot summer sun and hardened Bedu would probably have wilted, or more likely, would have avoided the location altogether. This audience of Ministers, Senior Officers, and VIP guests had long since lost any personal affinity with the Desert, as they now spent most of their time commanding desks in comfortably air-conditioned Offices, which is where they felt most at home. The heat proved just as hostile an environment for the audience as it did for us Visitors, and I sat back to watch the demo from the pleasant shade of my Exhibition marquee, enjoying a cooling breeze provided by a generator-powered-fan. I sipped an iced-Coke from my special box of iced-Cokes and watched events unfold.

The demo audience filled the unshaded seats and grew increasingly uncomfortable from waiting for the demonstration to start and from the increasing intensity of the sun. Delays to the arrival of the first target only heightened their already high anxiety levels. At last, an NDC Briefer addressed the audience from a priestly lectern placed before them. He was a northern 'Tecchie' in full suit and tie and spoke with a very strong regional accent, which was enough to baffle me, so quite what it did for the Arab audience was a mystery. Perspiration gushed from his every pore and streamed in torrents down his reddening face, he seemed to be melting before our very eyes; he spoke no Arabic and some of his audience spoke little or no English. There was no Translator.

When the first target finally appeared, a sense of anticipation mixed with an even stronger sense of relief rippled through the restless VIPs. On came the Drone, its engine tone getting louder and louder as it neared the engagement zone, and the Interceptor System twitched as it located and tracked its target, then swung to engage with its Command System locked on. The first Interceptor launched with an impressive bang and soared skywards in an arcing flightpath watched with great excitement by the sweating audience. The missile closed in on its target with increasing velocity and we all waited for the climax. It missed. Whoops!

The valiant Briefer improvised a hasty technical explanation, another Drone was on the way, just hold on to see the action. But it was too much for some. Officers in the rear seats glanced knowingly at each other and started to slip away and several of them found comfort and solace in my shady, cool exhibition Marquee, where I greeted them in Arabic and offered cold drinks, shade and chairs. They were very grateful, and we sat together and waited for the show to go on.

The uncushioned tiered seats in the sun were becoming embarrassingly empty, whilst my small tent was soon bursting at the seams with new Guests, who, ignoring the demo, inspected the display items with great interest and I arranged several appointments and ran out of business cards.

Several passes and misses later, the Demo ended and the melting Briefer bid thanks and farewell to the remaining VIP Guests and those unfortunate enough to be bound by seniority and diplomacy not to leave prematurely. The Briefer reminded them that they were all invited to an NDC Reception in a local luxury Hotel later that evening. I supervised the packing of my display Systems before adjourning to the hotel, and a rich sense of schadenfreude enveloped me. NDC's Glamour Marketing team slunk off, sweaty and depressed, to prepare for their evening hosting duties, whilst I went for a cooling swim.

I too had been invited to the Reception, though on a working basis to support NDC in a hosting capacity, although without a nominated VIP to sheepdog I would simply circulate and latch onto any promisingly available spare Generals or Ministers. Because of the Demo' disaster I guessed it would be a quiet evening and so it turned out. NDC personnel outnumbered the Guests as few of the senior Invitees turned up; they'd probably been casevaced with heatstroke. Company personnel ended up talking to each other as usual, quaffing copious quantities of their own, well Company-funded, Champagne. Tough life this marketing game, there's only so much Beluga a chap can eat.

Midway through the event an odious little NDC Marketeer appeared in front of me and silently pressed a note into my hand. He winked at me, and without saying a word disappeared back into the crowd. The note read 'Room 1625 at 2230hrs'. What was I to make of that? Had I just been propositioned by a gay Defence Marketeer or was this NDC mischief? Fatally curious as ever, I duly presented myself at Room 1625 at the appointed hour and rang the bell. Mr Odious opened the door and, pausing only to glance left and right along the corridor in classic Clouseau fashion, he ushered me in.

Inside the room was NDC's Regional Marketing Team led by that slimiest of slimeballs Jolyon Tennyson himself. In a different era, no doubt, young Mr Odious who'd opened the door would have been Tennyson's Public-School Fag, sycophantically delivering a full range of faggy personal services to his slimy Master; it was that sort of odd S&M rapport between the two of them.

"Have a drink…" said Tennyson, oozing over to me and offering his wet-fish handshake. "…you know these chaps I think…" Tennyson gestured towards the other two with his drink-free hand. I certainly did know 'these chaps'. In addition to Mr Odious, the Sorcerer's Apprentice, were two ex-military NDC Executives, discretely but insultingly nicknamed behind their backs 'Laurel' and 'Hardy' by NDC staff; Laurel was ex-Royal Air Force and Hardy was ex-Army. This dysfunctional pair were stuck in a bygone era and quite how NDC imagined that these two idiots would ever win any Defence business was a mystery to most of us. Tennyson invited me to sit down and Laurel and Hardy perched beside him on a sofa. They looked like the three unwise monkeys. Mr Odious sat on an upright chair to one side, presumably relegated to a flank because he was the junior member.

Tennyson explained that this cosy meeting wasn't purely a social affair, how could I ever have guessed, but rather an opportunity for us to share methods and pool regional information and contacts. But it wasn't just contact data which was a problem for them, they just didn't have the Business or Sales skills, the savvy, needed to win deals. They needed a 'Closer' and in their heads it was going to be me. Time was running out for this team, as back in UK their shortcomings were painfully exposed; they needed results quickly. It was an expensive team to maintain in one of the most expensive marketing regions in the world, and only high-value Contract-wins could ensure their continued employment. In any US Corporation, they would already be gone.

After token small-talk and regional chit-chat Tennyson got down to it and floated the idea of me joining them. Sales might have been the open topic but there was something else beneath the veil. Could they trust me with their grubby secrets, their kickback aspirations? That was the real question. They wanted to access my contacts and share the Arms Cash action which they automatically but wrongly assumed would be available, this was how they thought. I took it head-on as I was feeling punchy, it must have been the Champagne, and I told them exactly how I saw them, i.e. flailing around looking for Agents to do their own selling job for them, while their Competitors locked-on directly to Customers with better technology systems and won contracts.

Ever since the Gulf Wars everyone wanted American stuff; it was technologically 'hot', and you got the USA as a 'bestie' into the bargain; even bucket loads of Arms Cash couldn't overturn that proposition. Exceptions proved the rule of course, as with Hector, but in truth that was a legacy deal and not a road map to the future. The Defence world was changing technically and commercially and NDC had missed the key moves. They all looked a bit depressed, so I left them to it.

There was no immediate outcome that night, as too much free Champagne flowed, so I called Tennyson a few days later from Dubai and told him that I didn't want to work with them. He was very disappointed and asked me why not, so I told him that within a few days we would all get on each other's nerves. What I really meant was they would get on my nerves. Well, they already had. He told me they'd discussed it and thought it could work out very well and I could just picture them in a Hotel room together after a Contract signature, if they ever achieved one, dividing up the spoils and arguing over who got what; a Defence Industry take on Chaucer's Pardoner's Tale perhaps.

Some Defence Marketeers and Salesmen undoubtedly sought out and appointed only those Agents who were likely to come up with Arms Cash kickbacks, whilst others operated 'pick and mix' policies selecting both 'real' Agents and Kickback pals to spread their risks and opportunities. At the working level, such kickbacks would be modest, but at a senior level substantial rewards were realisable. It's certainly been the case that some really good potential Agents, those who may actually have been best placed to deliver business for a Company, were passed over for more compliant 'friends' who would deliver Kickbacks if they ever got lucky with an Order. In all these situations, ever shrouded in plotting, scheming and secrecy, absolutely nothing was as it seemed to be. The key was getting in close, very close. Everything was distorted in the Arms Cash Hall of Mirrors.

Chapter 3
Close encounters

Try to sell Arms to Iran today and at best you may be jailed or at worst terminated, depending on who catches you at it and what you are trying to sell, but it was not always so. My first close encounter with Iran came when I worked as a young Contracts Officer in an MoD Procurement team in St. Christopher House in London. This building was the largest Office Block in Europe when it was constructed, but has now been demolished; it was certainly the ugliest and most unpleasant of buildings in which to work, and my job there was to buy Armoured Fighting Vehicle items, including special equipment for Northern Ireland Operations. As I churned out my tedious Contract paperwork, I'd no idea just how close I was to a major Arms Cash deal in action, indeed, the whole concept of Arms Cash had yet to appear on my naïve young horizon.

St. Christopher House was also home to the MoD's 'Iranian Tank Office' which was located on a floor above my own Contracts Branch. The function of the Staff there, including the British Army Officer designated 'The Iranian Tank Officer' was to ensure the timely monthly despatch of British Chieftain Tanks to Iran from Marchwood Military Port in Hampshire. At that time the UK was sending around sixteen Chieftains a month to Iran, plus spares and mountains of ammunition, and c.900 tanks in all were ordered.

Arms exports to the Peacock throne were controversial even in those far-off days when Iran didn't threaten other countries with proxy attacks, because the Shah's autocratic and oppressive regime violated human rights. Iran sought to achieve regional military superiority over its neighbours, and specifically over Iraq, which if achieved, would destabilise the region, but Iran also needed military muscle to sustain internal suppression. As ever with British Foreign Policy though, Arms Export revenues trumped Human Rights, and Britain became the dominant supplier to the Shah and fed his voracious appetite for Arms.

It was truly a matter of sorrow that Iran, a once sophisticated and highly cultured Country, had become dominated by an insatiable quest for modern weapons. Ironically, Britain, also a once sophisticated and cultured country (?) has now developed an equally insatiable lust for making more and more Weapons to boost the flagging National Economy.

British Arms exports, and the associated deep in-country involvement in developing Iran's Defence-Industrial base, arguably created the blueprint for the later huge and even more controversial Al Yamamah deal in Saudi Arabia, and the common ingredients shared by these two deals are clear to see, viz. autocratic governmental clients, in-country human rights issues and the British preference for Arms revenues over ethics or morals, plus of course, Arms Cash.

The Iranian Arms business it was claimed as usual, represented an important contribution to the British Economy, but how many times have we heard that one? The 'Iranian Tank Officer' oversaw and reported on this lucrative Project, though no doubt he was unsighted on the Arms Cash elements. I was naïve about Arms Cash too then, and I didn't and couldn't see or even imagine what was going on behind the scenes, but an awful lot was going on.

To spare the Government's blushes in making direct Arms Cash payments with Taxpayers cash, the millions of pounds-worth of Arms Cash commissions for the Iranian deal were arranged covertly through a third-party cut-out Company called Millbank Technical Services, and British Ministers and Officials were at the very heart of these arrangements. The story is well-known and was well covered in an investigative report in The Guardian. The key point is that the Government was complicit in paying out Arms Cash. At the time of the Iran deal, the primary manufacturers of the exported equipment were the Royal Ordnance Factories which were still in Public ownership then, but which were later sold controversially and for a song to British Aerospace, now BAE Systems, which no longer has a majority British Shareholder.

The Government's 'Crown Agents' owned Millbank Technical Services which was used as the stand-off contracting party and as the vehicle for paying the Arms Cash. It was a classic example of a cut-out entity being used. Arms Cash 'Influencers' and 'Runners' were at work in the creation of this deal. The Shah's continuing and growing thirst for more and more British Defence equipment and for the deeper development of Iran's indigenous Arms manufacturing capabilities, coupled with the Middlemen's insatiable appetite for Arms Cash, were matched equally by the British Government's willingness both to supply the Arms and to pay-out the Arms Cash via secret channels.

It was envisaged at the time that Arms business with Iran could reach the £1-2bn value range, simply huge numbers back in the day and very distracting and seductive for Politicians. Ironically in 1993 a senior Ministry of Defence Official, Gordon Foxley, who was the UK Director of Ammunition Production, was himself convicted of accepting bribes from foreign Arms Companies. Foxley also worked in St. Christopher House not far from both the old Iranian Tank Office and my own Contracts office, although he was apparently unconnected to the earlier Iranian Arms Cash activity.

Thus on the one hand, the British MoD had been very happy to play a key part in managing the supply of Arms in a deal knowingly and covertly lubricated by Arms Cash millions, whilst on the other hand, when one of MoD's own Officials was caught receiving Arms Cash, he was jailed. Nice ethical selectivity. Perfidious Albion at work. The Iran-UK Arms Cash circus came crashing down when Ayatollah Khomeini and his supporters deposed the Shah; a lesson in succession which should be revisited quickly today.

Iran subsequently resisted an Iraqi invasion in a bloody and long-lasting war, initially using the Arms purchased from Britain; notably the Chieftain tanks. Although international Arms supply restrictions were applied to Iran during this period, Britain stayed ahead in cash terms, holding back some £450m of Iranian advance payments for Arms, mainly Tanks, which were never actually delivered. This unsatisfactory situation was only resolved decades later, apparently to ensure the release of Nazanin Zaghari-Ratcliffe, although her release could have been achieved much earlier with less fanfare and with less suffering. UK had held onto the Iranian money without justification for many years, and ironically the recipients of the Arms Cash thus benefited from commissions paid for unfulfilled Arms deliveries. Perhaps MoD should have demanded an Arms Cash refund; that would have been a first.

Once Saddam Hussein attacked Iran, ever perfidious Albion duly and predictably switched sides from supporting and supplying Iran with Arms, to supporting and supplying Iraq instead, although British export restrictions allowed the continued supply of non-lethal materiel to both sides during the early stages of the war. Nothing like covering your bets. Iraq became the favoured party in Western eyes and accelerated its own programme of indigenous Defence-industrial growth to counter the embargoes. At times it was hard to keep-up with the Embargo Merry-go-Round and remember which Country was embargoed by which other States.

Defence Company conversations in Middle Eastern hotels during the long Iran-Iraq War included thinly veiled references to the 'Right Eye' (Iran) or the 'Left Eye' (Iraq) as every Arms Cash Agent in the Region sought juicy Arms Cash opportunities with Defence companies which were ready to run the embargo gauntlet, or which were canny enough to find a work-around route. Once again, the penalties for getting caught-out could be severe, depending upon who did the catching, so the eleventh Commandment applied 'thou shalt not get caught...' The only sensible thing to do was to avoid both of them.

Comprehensive lists of the Defence needs-and-wants of both Countries circulated freely during the Conflict and continued to circulate after the Ceasefire. These lists included parts required to support items exported previously to either country by Britain and others before any embargoes were imposed, then released or re-imposed. 'Back-channels' to both countries remained both open and highly active. Third-party Countries which operated common equipment types were aggressively targeted as sources of spares, weapons and ammunition by the combatants. In a neat twist of fate Iran is currently undergoing a resurgence as a Defence manufacturer of Combat Drones which are being supplied to Russia for use against Ukraine. Lesson: The long-term implications of enabling any Customer country to become an independent Defence Manufacturer seem never to have been factored-in by British politicians, who should take note, before Ukraine is similarly empowered without controls. Oops, to late!

Britain supported Iran under the Shah when export revenues and politics aligned but switched rapidly to favour Iraq as the perceived threat from Iranian fundamentalists developed; it was not much later though that Saddam Hussein and Iraq became the enemy. Like the Iranians before them, the Iraqis worked hard to reduce their dependency on Western Defence Suppliers and an elaborate worldwide web of covert front companies procured and channelled Defence technology and know-how indirectly to Iraq. It was an Arms Cash paradise. At the infamous Baghdad *Exhibition for Peace and Prosperity* after the Iran-Iraq War in 1989, and prior to the Gulf War, Western observers were shocked to see the progress made by indigenous Iraqi arms programmes, although progress was not without penalty for some participants, notably for Dr. Gerald Bull who was responsible for the Iraqi Project Babylon 'Super-Gun' design. He was assassinated in Brussels in 1990 having met with a British Defence Company Executive shortly before his death.

On the international stage the pendulum of British Defence patronage swung from Iran to Iraq and thence to neither. In my little world, the proximity of the Iranian Tank Office in St. Christopher House had provided a first fleeting encounter with the Iranian Defence domain and Arms Cash, but my second encounter with Iran was to be a more personal, more hands-on affair.

The dry delights of commuting daily to Southwark in South London, to write Defence contracts for the MoD soon lost their allure, it really didn't take long. That the sole consolation for working there was being lunched by Defence contractors in the nearby 'Market Porter' pub, and in various Italian restaurants, was an indication of just how dull it was. I reached a Career morale low point as an MoD civil servant and opted instead to convert my military hobby as a Reserve Forces soldier into my full-time job for a while. I duly transferred into the Regular Army on a six-month attachment for operational duty in Northern Ireland. That was preferable to dull contracting duties in St. Christopher House. I didn't want a full Army career, heaven forbid, I'd seen what that looked like during my time in the MoD, I just wanted some military operational experience and a taste of adventure. My aim was to get to Northern Ireland, survive a tour of duty there, then move on to do something else. I duly transferred from the Reserve Forces into the Regular Army.

I served firstly at the School of Infantry and then in a Mechanized Infantry unit in Germany, from whence I finally deployed to Londonderry during a vintage year of violence during *The Troubles*. It seemed to me that many of the incidents which occurred there did so simply because soldiers were available to be shot at or bombed, and we referred to ourselves as the *'Duty Targets'*. It was beyond the wit of politicians to resolve the situation, so they ignored it instead, they were all asleep at the wheel, much like with Afghanistan later, so no change there. I got a bit carried away in uniform though, and overshot my intended six-month attachment by a further three years, and this time it was the drab, repetitive Cold-War routine in Germany on a restricted training budget which became the claustrophobic and repetitive factor. We didn't have enough ammunition, fuel or spares to train properly; we were deluding ourselves. I left the British Army of the Rhine in search of something more exotic, and after a Reserve Forces refresher, I headed off to the Sultan's Forces in Oman, where I joined an Omani Infantry Regiment in an operational area. I'd never been to the Middle East before.

At that time there were two categories of British Officer in Oman. Some were seconded from the UK Regular Army and others were 'Contract Officers'. Contract Officers were ex-British Army (and RAF and Royal Navy Officers) who'd been recruited and contracted directly into the Sultan's Forces via a small British company called Airwork Services. I well remember my Airwork interview in an office above Lillywhite's in Piccadilly in central London. It lasted all of ten minutes. The interviewing Officer thumbed through my military CV and asked me when I could start, and I was on a flight to Muscat two weeks later. Best interview ever.

Contract Officers were a mixed bunch. Some were left over from Oman's last Frontier Conflict and had been there for years, some liked to fantasise about themselves as Mercenaries, hoping to add a dash of glamour or mystery to their lives, but truly they were 'Club-class Mercenaries' at best. We all went to Oman for one of several reasons, typically including escape from a marriage, an obligation or a debt perhaps, or maybe for military adventurism in the mountainous and beautifully harsh terrain of Dhofar. Certainly, some were Gay officers seeking a less judgmental social environment, at a time when the British Army had yet to become fully 'inclusive', and finally, there were those who'd simply run out of options and had nowhere else to go. Make no mistake about it though, regardless of their individual motivations and characters, there were some highly professional British soldiers on contract to Sultan Qaboos, including several who really should have still been in the British Army. Of course, there were some complete 'Knackers' too, but the British Army has plenty of those also.

On arrival in Oman I was deployed immediately to an operational area in Dhofar Province on the Yemen border, where the bloody conflict with 'FLOSY' (Front for the Liberation of Occupied South Yemen) had ground to a halt a few years previously. Let's hope that Military planners who are undoubtedly creating Yemen land invasion contingency plans to 'neutralise' the Houthis, look back to this conflict for lessons, and learn too from Britain's ignominious exit from Aden in the '60s.

My Omani battalion patrolled the unstable and porous southern border with the PDRY and there were just two of us Brits in the unit. Dhofar Province is the starkest but most beguiling of regions, it is unique; beautiful and brutal in equal measure, a place which bites out a little piece of your heart and keeps it forever. Our Mine-strewn frontier positions occupied high ground way above the empty feral beaches which fringed the Arabian Sea and stretched inland to scorched and uninhabitable mountainous terrain; very arduous and great ambush country. A miraculous freak of nature, a seasonal monsoon called the '*Khareef*' visits this region annually and overnight the barren unyielding rocky landscape is transformed into a green wonderland. Sparkling streams bubble up as if from nowhere and greenery emerges overnight from the barest of rocks and sand. But the area was awash too with the debris of conflict and had been heavily mined and wired; there was a pot-pourri of lethal hazards both human and otherwise, and my second encounter with Iran was waiting for me there.

I'd returned from a border patrol and had deplaned from one of the ex-Vietnam Hueys operated by the Sultan's Air Force, when one of our soldiers, a '*Jundy*' in local parlance, hurried towards me shouting excitedly and cradling something in his arms. As he got closer, I could see that his prized trophy was a Mortar Round, defined by its bronze-green body colour and yellow banding as high-explosive; there was Arabic or similar lettering which I could not read on the casing. I urged the young soldier to stand still as I gently relieved him of the explosive round. The safety pin was missing from the impact fuse; it was a lethal item and I was holding it. I had the area evacuated, not too difficult in desolate Dhofar, then isolated the Round and set about preparing a plastic explosive (PE) demolition charge with fuse and detonator, just as I'd been trained to do at the School of Infantry some years earlier (thank you, Major McCurdy).

British Infantry Officers were, I suppose they still are, trained to destroy unexploded ammunition, 'Blinds', which occur during live-firing on Ranges and field-firing areas. I always hated having to destroy unexploded grenades, which seemed to swell up in my mind's eye as I walked towards them. I've never forgiven the Royal Ordnance

Factories for making so many lethal duds, presenting more of a danger to friendly forces than to any enemy. Luckily, I'd served in a Reserve Army Mortar Platoon for eighteen months, and although I'd hated every tedious second of it, that experience repaid me now in Oman as I prepared to destroy the unexploded munition.

After test-burning the fuse I moulded my PE Charge (it stinks) with my hands and placed it on the ground as close as possible to the Round without actually touching the Round itself, and took care to lay the fuse as straight and flat as possible, holding it in place with stones. I ignited the slant-cut end of the fuse with a special 'Match-Fuzee' then retired as directed by my Warminster training; you must walk away from the fizzing fuse, not run, to a safe position. Easy enough on a formal Grenade Range in the UK where a concrete Bunker is usually available for protection, but somewhat more difficult here, sheltering instead behind a bump in the ground in the Dhofar landscape. I rolled into cover behind some rocks and counted-down the fuse burn-seconds on my watch. The PE charge and Mortar Round detonated together, causing a significant bang in this naturally echoing bowl of a location, surrounded as it was by high rocky ridges. The Omani Southern Brigade HQ had previously alerted the Yemenis on the other side of the Border about the imminent detonation, lest they thought we were shelling them and retaliated.

My Omani sidekick advised me that the markings on the Round were Iranian, the writing was Farsi. It must have been left there by the Iranian Brigade which had been deployed to Dhofar by the Shah to support his friend, the Sultan of Oman, in resisting the Yemeni insurgency. Rich layers of irony here. That Mortar Round was of British manufacture which meant that the Royal Ordnance Factories made it; I worked with them when I was in the UK MoD. This Round was intended to help the Sultan defeat his enemies, but I was now the only one at risk from it, and when the Iranians helped Oman militarily in Dhofar, as did the British, Iran was Britain's friend. They subsequently became the enemy, but the British had already supplied both Countries, Iran and Oman, with Arms, including Chieftain Tanks, Mortars and ammunition manufactured by the Royal Ordnance Factories. A Royal flush of irony really! Perfidious Albion strikes again and again. The MoD Iranian Tank Contract in London may have been my first fleeting but unwitting encounter with the Iranian Defence domain, and this High Explosive Mortar Round in the searing heat of southern Oman was certainly my second, but a third and much livelier encounter was shaping-up.

When my Omani Battalion completed its operational tour on the Yemeni Border, the Soldiers rightly anticipated Leave, but unfortunately we were quickly re-deployed, this time to the Mussandam Peninsular, the tip of which creates the narrow strategically critical Straits of Hormuz separating Oman from Iran. The southern end of the Persian Gulf is barely twenty-one nautical miles wide at the narrowest point, and at that time Mussandam was protected only by a Company-sized Omani Army Infantry Unit of around 120 souls.

Intelligence Reports (don't we all look forward to them) indicated that the Iranians might try to seize the Omani side of Straits and close-off the Gulf to shipping; an action which would stifle the movement of most of the world's oil. Nothing much changes does it? Iran and Iraq were still at war with each other, so the seizure and closure of the Straits would enable Iran to strangle Iraqi supply channels. To achieve this decisive closure would mean landing Iranian troops in Mussandam which was Omani territory, so my Battalion was tasked to go to Mussandam and sit there just in case. We were beefed-up with additional tactical elements to form an ad hoc Battlegroup with Artillery, Milan Anti-tank missiles and other elements placed under Command. We deployed to Mussandam through a combination of land, sea and air delivery mechanisms, the logistics of which were planned by me on a blackboard.

We firstly flew Rifle Companies in C130s directly into the small Khassab airfield, then cross-loaded troops into Huey choppers which ferried troops to defensive positions high in the oven-temperature rocky terrain. We established surveillance over the deep Fjord-type features in Mussandam, it is Norway with extreme heat, a brutal, feral place to have to defend or even in which to simply sustain life. There was nothing there other than rocks and sand and sun and pissed-off soldiers who wanted their overdue Leave.

My fellow Brit' and I, just the two of us in the Battlegroup, speculated whimsically on the possible outcomes if the Iranians really did try something. We had this 'skinny' Infantry Battlegroup of c.900, which looked sparse when sprinkled over the rocky terrain of Mussandam. The Iranians could apparently hurl a whole Division or two, each numbering thousands of men, across the Straits if they so wished, supported by Air and Naval forces. It would all be a bit one-sided.

We mused that we could become the subjects of a revival of the great Victorian Military Art Print tradition which celebrated Imperial military annihilation in chromo-lithographic form for posterity; gallant and plucky, yes, but very dead. Iranian reconnaissance aircraft overflew our positions but nothing more aggressive occurred, and sometime later we were withdrawn.

Deploying to the Straits of Hormuz had been my latest close encounter with Iran, and the next one which was to follow a few years later also had roots in Oman. Like all the best 'Close Encounters' this one would involve actual contact. When my Oman Contract ended the Army Commander personally asked me to stay on and I was very honoured by his request, but sadly for me other factors were pressing in. I'd played with all the military toys in the Omani box and could never again enjoy such military freedom; sadly, it was time to move on.

I joined a small UK Defence Manufacturing Company to learn the Defence Sales and Marketing ropes, but the Interviews were tough because the Company was resistant to ex-Military candidates. This surprised me, but they'd had bad experiences with ex-Military types before and were nervous about getting caught out again. As an MoD Contracts Officer, I'd been on the receiving-end of various Sales and Marketing ploys and gambits; I'd been the Gamekeeper and now I wanted to poach; I told them I could spell the word 'Profit'. They hired me.

My first Defence Sales role was UK-based and some of my MoD Customers were in St. Christopher House, so almost home from home so Deja vu reigned. I also supported the Company's Export Sales teams which was much more exciting. Here I must digress to insert a note of gratitude to a dead friend. I reported to the General Sales Manager who was a hard-bitten, some might say cynical operator, an ex-Royal Navy 'Master Mariner' and 'Master Gunner', both titles appeared on his Business Card and looked rather impressive. Let's call him Jerry. On leaving the Royal Navy Jerry worked firstly for IBM, the 'Big Blue' before dropping into the Defence Sales world. Jerry had a highly focused approach to Defence business and sustained an incomparable and labyrinthine knowledge of Land, Sea and Air Systems. He constantly honed his own and his team's Sales skills using a blend of personal experience and external training provision, and his knowledge of international markets was second to none. In my experience he was unique in the Industry.

Jerry's golden gift to me was to surgically remove 'deference' from my mindset. He knew that former Military personnel carried deference to Senior Ranks and Officials into their Sales roles, it was ingrained in them by their Service training and they lugged it around like heavy baggage. This, Jerry explained, was limiting, so he de-programmed my military mindset and kicked out deference in the process. He didn't urge rudeness or cockiness, just the ambition, presence of mind, confidence and agility to engage at any level without batting an eyelid.

Jerry was way ahead of his time with his Sales ethos and techniques, but he'd endured a significant personal tragedy in his life and was later entrapped in an attempted TV Sting, not doing anything illegal. He committed suicide. Thanks to Jerry's sales training, military de-conditioning and 'ditch the deference' mantra, I've happily engaged with Ministers, Force Commanders and CEOs in many countries and at least two Prime Ministers during my Defence Sales life and I remain forever grateful to him.

I sold military products to the MoD initially, neatly ironic for an ex-MoD Contracts Officer, and it was Ok, and I enjoyed it for a while, but it was too easy, and I needed more, much more. Selling to the UK MoD is the kindergarten of the Defence Sales world and looking back at it now I don't even count it as Sales; Candy and Baby syndrome. I longed for international exposure, and after three years I left the Company to find it.

Soon I was running multiple business Projects, Prospects and Agents in the Middle East and Asia for a major Company. I did deals, lots of them, I liked it, I was good at it and I travelled to places other Executives didn't want to visit and happily worked alone. I enjoyed this international life, but ghosts were hovering in the wings.

By chance I encountered an ex-Oman Army acquaintance of mine at the Eurosatory Defence Show in Paris. Tony Green was an ex-British Army 'passed-over-Major' from a bygone age, who on discovering that civilian life was a bit of a struggle, made his way instead to Oman, where thanks to an Old-Boy network connection he was hired by the Sultan's Forces. We were both serving in the Oman Army when I first met him in an Officer's Mess in Muscat during one of my rare visits to the Capital from my remote Infantry Base.

Green was a sad character who'd struggled to adapt to life outside the British Army. He missed the comfortable predictability of it all, the Mess life, the antiquated traditions, the Port and Stilton and the structured deference. His subsequent expensive divorce highlighted his inability to come to terms with his new civilian life or to earn a living, so when Green discovered the Oman Army option he was in his element. Here were all the familiar benefits and more of a military life, with none of the career, financial or social pressures with which he'd struggled at home.

Green and I were both 'Contract Officers' who didn't serve in the Oman Forces for a career. A standard Contract term was two years, though many Officers extended their service a year at a time to prolong the agony of having to go home again, and Green did exactly that. He'd been in Oman for six years when I met him there. Green was born in the twentieth century, but his personal and social expectations and his emotional intelligence too, were rooted in an earlier age. I mean it kindly when I say that he was a time-displaced misfit, but he was not alone as such in Oman, which seemed to attract them.

Green knew he'd never have reached high rank in the British Army, though in the good old days this didn't matter as the Army would have found him a post somewhere to see out his time. In the new career-driven Army this was no longer realisable, and for all his blind and unquestioning loyalty, Green was rejected by the Army he'd served so loyally.

In Oman none of that mattered. No one was going to judge him and there were no competitive Career or social pressures to manage. The Omanis paid him tax-free and in his remote Camp he was the 'King' and he led an easy-going, pressure-free life of Raj-like quality. He was liked well enough by the Omanis and was waited-on hand, foot and finger by a Platoon of Asian servants in an Officer's Mess inspired by the Raj. It all fitted well with Green's perception of his own status as an Army Officer. Somerset Maugham could have written a sublime short story about Green and his life; he was a Maugham character through and through.

As he approached the end of his much-extended Contract, reality began to gnaw at him again. Green's divorce had cost him dearly and after years in Oman he felt out of touch and vulnerable to the future and to old age and he started to panic. Green manoeuvred himself away from his role in an outlying Unit into a minor Staff job in Muscat, hoping to find new work opportunities in the Capital to replace his Army role, but this move was to prove his undoing. Maybe it was fear for his financial future which motivated him, but Green was seduced by all the chatter he heard in Hotel Bars about Defence deals in the Sultanate. He met visiting Defence Salesmen socially in Muscat hotels and became convinced that he could get-rich-quick by using and selling his local Insider Defence knowledge for Arms Cash.

Green was drawn into an ill-advised proposition and became a Pawn in an Arms Cash deal in which he was massively out of his depth. In his naivety, he badly misjudged the senior Omani Officer whom he'd hoped to engage in his secret Arms Cash deal and the plan backfired. The Officer took personal and moral offence to Green's proposition and reported him immediately, so after years of military service in Oman, Green was given twenty-four hours to leave the country and locally it was regarded as a generous let-off. He might have been jailed and publicly disgraced for corruption.

Once back in the UK, Green fumbled around again looking for a job before the 'Old Boy Net' rescued him once more, this time getting him into a UK Defence Company as a Far East Salesman. Green had no sales credentials and no knowledge of the region either, and predictably managed to sell nothing for the Company, but this international role satisfied his ego with its jet-setting ambience and luxury hotels.

Green had first heard Arms Cash stories at the Customer-end of the spectrum in Oman, and now he was hearing the same stories at the Supplier-end of the spectrum, so he focused on the potential for the lucrative kickbacks which were rumoured to be paid to some international Salesmen. Like urban myths though, these rumours travelled well ahead of reality, they occurred of course, but not with the ease or frequency which Green and many others imagined. He also obsessed with the potential to freelance whilst overseas by setting up little deals of his own with other Companies at his own employer's expense. Just how to exploit all these great Arms Cash opportunities remained his challenge though, because he lacked the essential business savvy, and his negative experience in Oman made him cautious. Never one to do anything on his own Green looked around for a Rainmaker, and sure enough one hove into view.

During his Oman Army service in Muscat, Green met a charismatic Iranian businesswoman called Fatima; quite how they met is a mystery since two more different characters would be hard to find. Fatima was commercially savvy; she was smart, sassy and internationally well-connected, whilst Green was a donkey. Green hadn't realised it at the time but behind the scenes Fatima had masterminded the failed sting in Muscat which 'did for him'. When the deal flopped-and-dropped like a lead balloon it was Green who was caught and thrown out of the country, whilst Fatima, who pulled all the strings but never appeared above the parapet, simply melted back into the international mist unscathed and as anonymous as ever, as was her well-practised way. Fatima was never one to waste a contact though, however feeble, and she ensured that the two of them kept in touch from time to time which fed Green's flagging self-esteem. Green was putty in Fatima's hands; well, most men were. The lady was as charismatic and glamorous as she was clever and dangerous, trust me, I know.

Like a child eager to show off a new toy, Green called Fatima when he landed his Defence Sales job and the chic Iranian flew into London from Paris to meet up with him and celebrate his change of fortune. Fatima and Green were reunited over a very expensive lunch for which Green paid. Green, the middle-aged, divorced, desperate, broke ex-Army Officer always paid when he met up with her. Fatima knew from her Oman deal disappointment that Green was an unreliable operator, but he had his uses, and what he had now were contacts inside major UK Defence Companies and the UK Defence Export Organisation. Such contacts had significant potential value to Fatima, as the Lady was shaping new ventures and deep Defence Industry contacts were exactly what she needed next. Green and I were working for different British-based Defence Companies when we met up by chance at the Eurosatory Defence show in Paris.

Defence Exhibitions have become an Industry within an Industry, primarily benefiting the Organisers, and these shows absorb huge chunks of Defence Company resources. Actual deals are generally done elsewhere and away from all the ballyhoo, although deal-signing is sometimes staged to project national or corporate exhibitionism. Many Senior Execs, usually those without responsibility for selling anything, love these Corporate cock-of-the-walk exercises, as do Peace Activists. Having said that, I've enjoyed, if 'enjoyed' is the right word, some entertainingly quirky moments at Defence & Aerospace Shows around the Globe, although I wouldn't hurry back to one now. I had several appointments with senior overseas industrial visitors, 'SIVS' in Defence Industry parlance, at this Show, whilst Tony Green was stuck with unenviable Stand-manning duties for his new Employers. His thankless task was explaining Defence equipment Exhibits to casually passing non-Buyers, i.e. Press and data-collecting Delegations, and we met there quite by chance. Because catch-up time at the Show was limited, we agreed to meet for a drink later in central Paris, for old time's sake.

Frankly speaking, I usually avoid ex-Army contacts and never attend Regimental reunions etc., which are by nature exclusively and depressingly retrospective. 'Do you remember when we…' etc., etc., the ever-unchanging reunion script extols times and events past, which glow ever rosier as they recede more deeply into the thickening mists of time. Old Operations, near misses, dead friends and so on, they pass like wispy holograms. Why recreate such trauma? The once common bonds become thinner and stretched with time and can only be refreshed amongst the Needy with regular and constant repetition; Reunions. It's unhealthy. Better to look and plan forward and not become defined by questionable memories.

Green and I had barely sipped our drinks in the Parisian rendezvous Bar when we were joined 'unexpectedly' according to Green, by Fatima, who just '…happened to be passing through…' the same Hotel Bar at the same time. Really? Green hoped that either me or Fatima or both of us would create and stage-manage some huge Arms Cash deal from which he would receive a life-enhancing kickback. At the time I was blissfully ignorant of the inside-track in the Green-Fatima-Oman story, I only discovered that later. I knew through grapevine that Green had been thrown-out of Oman, but I didn't know why, a Gay indiscretion perhaps, and I didn't know either that Fatima was behind the bodged Muscat Arms Cash project. Fatima didn't linger over her drink; she was vetting me, and she was flirting. A 'Runner' I wondered, a high-class Runner? She had all the hallmarks except gender, which made a pleasant change.

Fatima contacted me in UK a couple of weeks later and invited me to a meeting in London with her Svengali-like business partner, an Iranian Professor she referred to as 'The Doctor'. Quite why so many of these Agent, Runner, Fixer characters call themselves Professor or

Doctor I really don't know, but I've met several thus titled and heard of even more. I'd never have entertained any scheme devised by Green alone, because he was inept, but the Doctor and Fatima were in an altogether different league it seemed, so now it was my turn to be the moth drawn-in by the dancing flame. Was there some deal to be done here?

In the mirrored halls of Defence Sales and Arms Cash, the combination of feckless optimism and opportunism can have fatal results, literally, but it didn't follow that simply because Fatima and the Doctor were Iranian, they were about to propose some illicit embargo-busting deal. In any case, any Defence Salesman worth his salt draws from a deep Well of international contacts and you don't have to do business with everyone you meet. The currency of any and every contact is limited, it's that Gen. Tommy Franks dictum again '...*you have to trade...*' People move around and spheres of influence and touchpoints change and evolve all the time, some blossoming, others withering on the vine, just like relationships between nations and lovers. Iran and Iraq fought each other to a bloody standstill but still didn't sign a ceasefire; in the West the Iranian Fatwa demanding Salman Rushdie's death still echoed. Where did these two characters fit into the picture?

International events are frequently sudden and traumatic in nature; the Ukraine invasion for example, or 9/11, Gaza, Iraq, Afghanistan etc., and the impact of conflicts on the Defence Industry can be immediate. Some Customer countries, or aggressors and their allies, can become taboo overnight, whilst others suddenly become best friends. Western Arms-supplying States and Defence Corporations (in some cases it's getting harder to distinguish between these) are notionally restricted by National Arms export legislation which varies from country to country. If your Customer Nation doesn't fit the export template, then no Arms are coming your way; that's how it's supposed to work anyway.

The current political appetite for 'Arms Gifting' on a massive scale, such as to Ukraine, has disturbed the old order, so in some cases no marketing effort at all is required now to sell stuff, as inept Politicians are giving weapons away free. Easy. Wonderfully, there's no need to worry about the Customer paying for this stuff either, as your Government has decided that domestic taxpayers will pay. Did you vote for that? In such circumstances foreign Suppliers don't even have to pay-out Arms Cash, though I bet some changes hands anyway. There have been over thirty wars and conflicts in the twenty first century to date, so all that Defence Industry soul-searching after the Fall of the Wall, and that heart-rending claptrap about turning swords into ploughshares is long gone; these are the new Arms Cash glory days. But Arms Cash glory is transient. Iran was Britain's most lucrative Defence sales customer once but became an embargoed pariah State virtually overnight, leaving Iraq to become top-dog, until Saddam Hussein slipped-up over Kuwait. Who is to say that the political wheel will not turn again and reverse the Ukrainian, Israeli, neo-Iranian or other geo-political Arms situations? Play with fire; burn.

With Fatima and the Doctor, it was the same as any other contact situation. The key is always judging when to quit and break-off and when to push on. Did they really have some golden 'in' to a deal somewhere? Optimism keeps you moving forwards. The Holy Grail was ever to identify or to create a readily and legally (!) convertible international Defence Sales opportunity with minimal competitive interference, none if that could be arranged, and certainly without any Tendering nonsense. Today's Political Arms Gifts meet these criteria perfectly.

If the Lady and the Doctor really could do business somewhere or other, it wasn't obvious to me, and I really couldn't suss out into which category they fell, Runners, optimists, semi-discreet Government representatives, cut-outs or just Games-players? There was no outward hint of anything to do with Iran or any other country, or even any Defence equipment for that matter. Meeting them was interesting, yes, you can never have enough Contacts, but it could be just a damp squib without any content. Too early perhaps. Nurture a contact and maybe a gem will appear eventually, otherwise it's onto the backburner and move on. A couple of coffees later I was out of there and gone. I thought it was going to be a Runner conversation, but maybe I'd misjudged it. In this business you get to kiss a lot of frogs, but at least in Fatima's case, a glamorous frog and that made a change.

Two weeks later, a Dinner Party Invitation arrived out of the blue from Fatima. I'm not a Dinner Party person but if it's a forum for business dialogue then it's different, count me in. My best deals have been done in small discreet get-togethers, not at Dinner Parties. One-on-one is the best formula, although in today's compliance-pissed

Western corporate world, this is virtually impossible, and some Companies would rather lose a contract than transgress their self-imposed Compliance and Risk protocols.

As an aside, and prior to Fatima's invitation, the last Dinner Party I'd attended (in Chelsea) had been rather good. Maybe it was the exception which proved my dislike-of-Dinner-Party rule. The Hostess was the richest woman I know, an exciting elegant, sexy Jewish-Irish West-End self-styled 'Landlady' who rented-out Apartments in Knightsbridge. She was the former philosophy student of a Publisher friend of mine and I met her through him at her Chelsea home. Her partner at the time was a hard-headed Liverpudlian international bulldozer Salesman, whilst Hannah, my fellow Dinner Party guest, was a Jewish concert pianist. Since I was hot-foot, metaphorically and almost literally from service in an Arab Army, this wonderful combination of backgrounds made for a sparky evening. Rare. Would that be the case with Fatima's party I wondered? I doubted it; too hard-headed I thought.

I presented myself at Fatima's apartment in the London Queen's Gate Embassy zone; if this was going to be just a social thing it would be a chore, and therefore brief as far as I was concerned, although I didn't really believe this Lady would ever throw a Dinner Party just for the fun of it. I jogged up to the first-floor entrance and pressed the bell-push and I was shocked when the door opened. I'd met Fatima only twice before, once in a Paris bar supposedly by chance thanks to Tony Green's arrangements, then later in a London Office with her Professor friend or partner or whatever he was. I'd only seen Fatima in her City clothes, expensive and chic yes, but functional working gear for her, nevertheless.

Standing in the doorway now was an altogether different vision, for here was a supremely glamorous international lady and that expensive 'old-money' gown oozed style. Clusters of stones sparkled their sparkle and added a splash and dash of Middle Eastern bling. After a double-cheek 'Euro-kiss' Fatima ushered me inside; Green was there already there, and so was the Doctor who greeted me in Farsi. The other Guests were male, all soberly suited in smart dark, almost black business suits with the tieless Khamis fully buttoned; Iranians of course. All the alarm bells went off at once in my head, but there again, who dares...

The chatter was unremittingly geo-political, so good, no trivia, but I had neither the heart nor the nerve to tell them that I'd once occupied a rocky outcrop in the Mussandam Peninsula waiting, albeit with minimal military resources, to engage the Iranian Army had it leapt across the Straits of Hormuz. They'd probably have enjoyed the irony of the story. Neither did I relate the tale of the unexploded Iranian Mortar Round, which I'd destroyed in desolate Dhofar Province, nor that I'd once worked in the same Department as the British Army 'Iranian Tank Officer'. Iran had been Britain's friend once, but not now as perfidious Albion was at work. The cycles of modern history are fickle and rotate quickly.

The prevailing Western image of Iran at that time, featured radical black-clad Mullahs hell-bent on the total-destruction of anything Western, and especially of anything American. The tragic bloodletting of Ayatollah Khomeini's Revolution and the Iran-Iraq War obliterated the history of Persia's sophisticated ancient culture, supplanting it instead with images of US hostages and Iran's subsequent path to State sponsorship of terrorism. The Iranian regime regressed beyond political recemption, and the nightmare of the 1981 Iran-US Hostage crisis embittered Americans for all time. Many US Defence Industry Executive I met, even many years after these events, could hardly wait to bomb Iran. They probably will at some stage. Successive post-Suez generations of British Politicians have ridden along lamely on America's coat tails ever since 'Operation Musketeer' failed. They lack the intellectual capacity to engage creatively in the quest for regional stability in the Middle East and there remains a complete lack of vision, with no ability to innovate, originate or to create solutions.

Interesting though the Dinner Party chatter was, it finally gave way to something else.
"We need your help…" came the plea "…we need some things…" and thus, we got to the nub of the evening. Battle damage and attrition had reduced the Iranian tank fleet, and ammunition stockpiles had been consumed. Spares for most key systems were long exhausted so replenishment was urgent but legally impossible owing to the international embargo situation. Iranian Runners were foraging the world for military materiel to replenish stocks, encouraged initially by predictably ambivalent British foreign policy in the early stages of the

conflict, and manifested practically in the duplicitous activities of MoD's former Arms Supplying 'cut-out' entity, *International Military Services (IMS) Limited*, whose antics were well described in an April 2010 article in *The Independent*. UK Arms supply policy had changed as the war intensified, but both Iranian and Iraqi Defence equipment shopping-lists circulated around the world. Defence Companies were approached on a regular basis by Runners, other intermediaries, and Chancers of all sorts, hoping to do lucrative business for Arms Cash rewards.

Unluckily for these Runners, the British Chieftain Tank had been an export flop. There were some in the Region, but these had been political sales rather than operational necessities. Jordan (which later received Tanks diverted from the frozen Iranian Contract) plus Oman and Kuwait, operated a few Chieftains, but these were not major fleets. Political expediency mitigated against these States overtly supporting either of the combatants (well, too much, anyway!) and in any case they needed to conserve their national Defence inventories in case the Iran-Iraq war blossomed into a broader regional conflict.

Without a deep export pool from which spares or ammunition could be obtained, Iranian buyers faced the challenge of engaging covertly with the original manufacturers, the recently privatised Royal Ordnance Factories, which had been controversially sold for a song to British Aerospace. The Ammunition was by far the most difficult commodity for the Iranians to obtain, as many spares could be reverse engineered locally.

The ammunition was quirky and technically difficult to clone, and no other Nation used the same type of 120mm rifled Tank gun as Chieftain. It was a British Army 'doctrine thing' to keep this weapon in Service, because the Chieftain's gun and the ammunition were supposedly optimised to defeat the ever-evolving Soviet T-Series Tanks, which the British Army of the Rhine expected to have to engage one day.

The Cold War's long anticipated Nord Rhein Westphalia battle never materialized though, but a handful of Chieftain's successor Tanks, 'Challengers', finally played out the 120mm gun-versus-T-Series tank duel, in the Kuwaiti desert against Saddam Hussein's Army and the scenario was repeated when Challengers were gifted to Ukraine. A neat irony.

During my own service as a Battlegroup Operations Officer with the British Army of the Rhine (BAOR), and when the East-West German divide still existed, one of my key tasks was to covertly visit my Unit's assigned deployment area on the inner German border to assess potential combat positions. British Tanks needed to kill Soviet Tanks in a calculated ratio to halt a Soviet offensive, even though the Chieftain/Challenger ammunition enjoyed zero interoperability with any other NATO nation. So much for NATO's standardization and interoperability mantras; Ukrainian Forces will face the same challenge. From a Field logistics perspective, I don't envy the Ukrainians, it's an operational nightmare and they will probably dump the surviving tanks at some stage when the ammunition and support runs out.

I empathised with these Iranian buyers from a military perspective because no one wants to run out of ammo', but even if I wanted to risk my health by helping them there was simply no access to a source of supply, and doubtless, they already knew that. It wasn't a matter of being unwilling to help them, no one could help them with this. No amount of Arms Cash could make any difference, which was a bit of a pity, since Royal Ordnance were famously generous Arms Cash payers.

We talked about other stuff and parted on good terms. Desperation had driven them to explore every possibility and for my part I understood that. With UK's well-established reputation for changing its mind, and changing sides at the drop of a hat, it was entirely possible that I'd meet them again one day and be able to do business, so I was always going to be as friendly as possible. Who knows, we might even get a British Foreign Secretary one day who actually understands international affairs; I know it's just a fantasy, but I live in hope.

Fatima remained quiet throughout the evening as her primary job had been to get us all in the same room. She was a very accomplished Hostess, and I guessed she had influence way beyond this task; the clues were there. The knowing interaction and nods between her and the Runners, the glances, the eye contact. I was the last Guest to leave and Fatima implored me to help them if I could, and I assured her that I would if it became possible, but not at any price. It was more than a euro twin-cheek kiss with which she bade me farewell, and I was on a 'plane shortly afterwards heading to Dubai on a very different assignment.

The Iranian Radicals-of-the-day regarded Dubai as a modern Sodom and Gomorrah, but whilst the Iran-Iraq war raged it became commonplace to be contacted in Dubai by Iranian Arms Buyers. Dubai became the 'go-to' place for Arms Dealers to meet. Sometimes I'd barely checked into a hotel when the room 'phone would ring and someone claiming to be 'in the market' would ask to meet me. If this happens too often you need to keep on the move. Stick to the Military adage; never set patterns or establish predictable routines.

Tony Green called me after I'd been in UAE for about a week and urged me to get to a secure comms link and call him back. He wanted to send me a document by Fax. Fax? Yes, it had to be Fax he said. It was all very curious, what could this be about? I went to the 'inner office' facility in my Hotel Club Suite which had a Fax machine and I called Green back on a neutral 'phone. I treated no comms in the Region as truly secure, but this was as close as I was going to get. Military Signals training had drummed into me that three parties are always involved in any form of electronic communication: The Sender, the Receiver, and the Interceptor. Good advice to carry around.

Green faxed through some papers which I took to my room to decipher. After the unfruitful London Dinner Party, he'd continued to work with Fatima on other propositions for the Iranians I'd met in Fatima's Apartment. Tank Spares and Ammunition had slipped completely from their agenda because they were impossible to source, so now Green and Fatima were apparently firmly focused on Fast Patrol Boats. The Iranians needed to deploy more of these around the various Islands in the Shatt Al Arab waterway. This was nothing new, but the Iranian Navy needed not simply more boats, but more capable and better equipped boats and it was these new specification Boats which were the subject of Green's encoded faxes.

Fatima had apparently sourced suitable fast-patrol boat hulls somewhere in Eastern Europe and had opened her communication channel with Iran to discuss the price and availability of these craft, with a view to purchasing them in a back-to-back deal on behalf of her Buyers. As part of the deal the boats would have to be re-equipped and upgraded with modern weapons and systems, so Fatima also set up contacts in Asia somewhere, where she hoped to place the fitting-out and commissioning work. Green concentrated on sourcing equipment and worked on the logistics of getting the fit-out items to where they were needed. The Arms Cash commission and kickback potential from this deal was huge and it was exactly the sort of deal about which Tony Green must have fantasised about night after night after night.

As usual though, there was a catch, and it was a very big one. Someone had to go to Tehran in person to meet the Iranian Principals who'd been represented by the London Runners. The T's and C's of the deal had to be agreed and technical specifications for the vessels confirmed in a contract document. The financials also had to be negotiated and agreed, and for some unspecified reason this had to be done very quickly. Knowing that I was in UAE, Green suggested to Fatima that he could persuade me to go to Tehran on her behalf and get this done!

Fatima probably had good connections in Tehran, and the people I'd met in London were realistic about Iran's situation, but that was in London over dinner, Tehran was a cosmic step further away. Cleverer people than Fatima and Tony Green had tried and failed to get Defence equipment into Iran, and a flotilla of newly equipped Fast Patrol Boats was going to take some concealing. It had the fatal air of fantasy and delusion about it.

Green started to explain on the 'phone in 'veiled speech' (the Army loves that term) what they needed me to do for them, but I cut him short. I took a Taxi to another Hotel and re-established contact and I asked Green how much they proposed to pay for undertaking this suicidal mission, if indeed anyone was stupid enough to agree to go, and do whatever it was that was required. He said they'd yet to agree on a figure.

"You don't seem to have any of this worked out…" I said. I'd no intention of getting back to him, but conscious that Dubai was awash with Iranians I opted to tread carefully by not sounding completely dismissive. Green called me back three days later and had just started to talk about money, Arms Cash, when I interrupted him.

"Haven't you heard the News this morning?" I asked him. He had not. Ayatollah Khomeini's infamous Fatwa against Salman Rushdie was back in the news. Rushdie apologized for offending Muslims in 1990, but Iran had not accepted the apology and the Fatwa continued. In 1997 the bounty was raised to $600,000 and in 2000 was raised again, this time to $2.8m. In 2006 it was stated by the Iranian Martyrs Foundation that *'The fatwah by Imam Khomeini regarding the apostate Salman*

Rushdie will be in effect forever'. Rushdie was eventually attacked and seriously wounded by stabbing in 2022.

The book which made Rushdie a pile of money also resulted in violence, riots, bombings, and deaths around the world, and at the time of Green's call to me it was in the news in the Middle East. What Western Defence Company Executive in his right mind, or even in his half-right mind, would be making a solo covert no-questions-asked unsupported illicit trip to Tehran to discuss an Arms Transfer?

"It'll be OK…" urged Green desperately "…Fatima will fix it…" I considered Green's credibility and personal track record for no more than a nano-second.

"You're a complete idiot. F*** off" I said and slammed down the 'phone. For the first time it was a relief to get out of Dubai and I never again saw Fatima, the Iranian commercial femme fatale, nor Tony Green, who died an early death a few years later. I never made it to Tehran, which in one respect is a pity, since I was curious about ancient Persian culture and Cyrus the Great's Empire, whilst in another respect it was good to still be breathing. Omar Khayyam put it succinctly *'The moving finger writes, and having writ moves on to write again….'* I moved on too and took flight to the mysterious East.

Chapter 4
The Jewel in the Crown

For years I'd longed to visit the great Indian sub-continent and I'd planned a trip there once whilst on Leave from the Oman Army. The Iranian threat to the Straits of Hormuz intervened though, so I was deployed instead to the oven-hot heights of Mussandam with my Battlegroup. I was out of the Military and working for a major Defence Company when the next India visit opportunity came my way, and this time I was paid to go in style.

Defence Company travel policies vary in the level of Airline Cabin class which are allocated to their international Business travellers, and whether by accident, design or just incompetence, I was booked out and back to Indira Ghandi airport on a British Airways First Class ticket. I'd anticipated Club Class, and that would have been fine; heaven only knows what these First-Class tickets cost. I didn't argue the toss over it, to coin a phrase '…because you're worth it…'

Cabin class issues might seem trivial to non-travellers, childish even, but the Class of cabin is a critically important issue to travelling Defence Executives, especially to those who travel frequently and far. Aside from the vital primary issues of ego and Air Miles accumulation, there's a genuine need for Executives to arrive in good shape so they can hit the ground running (do any really do that? No) and Company travellers hope that their Employers genuinely value their health and welfare. They don't of course. My experience has been that US Defence Companies are more reluctant to choose comfortable air travel for their Personnel than are their European equivalents, although one US Company I worked for wouldn't even let me drive to or from a home Airport. Instead they insisted on sending a car to collect and deliver me each time. It was 'risk-mitigation' of course as they didn't want to be sued for the death of a jet-lagged Executive driving 100 miles home from an airport. Comically though, their preferred car company consisted of just two Owner-Drivers who drove seriously fast 'R' type Jaguars at Mach 1 speed up and down the M3 motorway in the wee-small hours of the morning, or late at night, so I really would have been safer driving myself, jet-lagged and half-asleep or not.

Airlines vary massively in their interpretation of the quality of different Cabin classes. Air Zimbabwe or Egypt Air for example did not compare even remotely with say, gorgeous Singapore or Emirates Airlines. As a rule of thumb, intra-European travel in Economy Class was just about manageable if not ideal, whilst the consolation prize 'World Traveller' type intermediate Class might suffice for slightly longer journeys. 'Club' really should be the standard for everything else. Predictably some Executives make a pure science out of their travel arrangements, to the point of obsession, but having experienced both the high and low points on the air travel spectrum, I am content to simply be 'comfortable'. I've squeezed in between sweaty, flabby returning tourists on long-haul Economy, I've travelled alone in Asia in eerily empty Club cabins, I've been bounced around in the back of low-flying Antonovs in Eastern Europe, I've flown in pampered luxury First Class and I've been deafened in a ramp-down Military C130 flying over the burning Kuwaiti Oilfields. I've also endured the ultimate mind-wrecking boredom and ultimate discomfort of pre-contractorised military flights to and from the UK to the Falklands Islands. I left a Defence Company once which kept booking me out on EasyJet's low-cost, low-amenity tickets.

It pays to be philosophical about air travel so being 'comfortable' is generally good enough. That flight to Delhi though was serene, and the First-Class cabin was virtually empty, there being only two other passengers, so we each received more than our fair share of attention, food and alcohol. One ex-army Officer Defence Industry associate of mine, expressed his strategy for managing long-haul flights thus: 'Drink as much wine and booze as you can manage in the first sixty minutes, then crash (i.e. sleep) for the rest of the flight. You won't feel a thing!' I can testify that he was very good at this, he certainly practised what he preached. Trained at home, I think.

Indira Ghandi airport was choked and chaotic, with passenger bodies pressing closely against each other and against me from every direction; spatial concepts there were very different, and newcomers found it uncomfortable. I took a black and yellow Austin Ambassador cab from the airport, and after witnessing my first three Indian traffic accidents during the first few minutes of the drive, below average I believe, I arrived at the New Delhi Oberoi Hotel, the Raj-style one with mahogany panelling.

Liveried doormen hastened to open the cab door and released my luggage from the boot; I was saluted by doormen as I approached the building, rather smart salutes in fact, and soon I was enveloped by the Oberoi's lush, panelled interior, an homage to Raj design. Why did they do that I wondered? The rooms were comfortable with a gentle, slightly facing grandeur and a Butler presented himself at my room and offered to unpack my luggage. Downstairs, the adequately equipped but empty Gym, with chrome plated weight-training equipment and a big outdoor pool, were just what I needed.

Aside from a recreational recce of the hotel Gym & Pool I made my usual assessment of exit routes and entry points. I walked outside the building to look back at it and considered how I might approach it if I was a Terrorist. Following the 2008 Mumbai terrorist episode which involved the Oberoi Trident and the Taj Mahal Palace & Tower hotels, I take more care than ever, since in those incidents Terrorists went from room to room looking for and killing Westerners. Survival and exit plans are required for every Hotel, and you must make a room-blocking plan.

With my security preliminaries completed, I took the first of many swims in the outdoor pool to the accompaniment of melancholy calls and wails from the Peacocks which strutted around the garden surrounding the pool. Their calls drifted up hauntingly through the balmy evening air and into the black but starlit evening sky. Would the business delights of this trip match the natural ones I wondered? I swam up and down alone in the pool thinking about what I might achieve in this fast-developing crazy country.

Getting anything out of this first trip would probably be impossible and my first task was really to start strategising. Where did the opportunities and touchpoints lay and how should I approach them? Were there actually any opportunities to be approached? To assist me in this process I was mandated to interface with our resident Corporate Marketeer who managed the New Delhi Office; I hadn't met him before. After my swim, followed by a spicy Dinner, I reflected that it was all going rather well so far in India and I sat back in a comfy Wing chair to take in the latest news from '*The Times of India*'. There it was all over the front page again.

The Bofors Arms Cash scandal had resurfaced, as it did periodically, and the Mother of Arms Cash deals was a headline yet again, with some new claim or development. It was never going to go away. It was way back in 1986 that the Indian Government signed a huge Howitzer contract with Swedish Arms Company Bofors, for the

101

supply of four hundred 155mm guns at the simply gigantic cost then of around $1.4 billion dollars. This procurement decision apparently went against the wishes of the Army, as a French Howitzer had performed better during the Trials. A year or so later, and bizarrely, following claims made on a Swedish radio programme, it was alleged that Arms Cash kickbacks had been paid to secure the deal. Shock horror! Surely not?

This initial revelation triggered a series of investigations and probes with various Swiss and other Bank accounts being frozen and unfrozen, and middlemen being identified and hunted down internationally from Dubai to Argentina and Malaysia. Arrest warrants were issued and quashed and issued again. Rumours of involvement and complicity ran to the very top of the Indian Government, with Rajiv Ghandi, the Prime Minister, being variously accused and cleared and accused again of involvement. The scandal ran for twenty-five years and impacted severely on Indian politics. Industry Insiders used to refer to 'Bank Account Lotus' in Switzerland as the safe harbour for some of the illicit Arms Cash, but Accounts were also discovered in London and frozen, then unfrozen.

It was the Arms Cash deal which had everything; persuasive Runners, a 'Mr Big' style fixer, Politicians, foreign suppliers, huge payoffs, shock revelations and premature deaths. Even the much-accused Italian businessman at the heart of the controversy had a wonderfully exotic name, one for which most novelists might give their all, Ottavia Quattrocchi, and even his earthier business connections with an Italian Fertilizer Company could not lessen the exotic nature of it all. It was a heady time to be in India on Defence business and Official eyes followed my every move, but despite a ban on using Agents, potential Agents queued up, now even more aware than ever before, of the scale of potential Arms Cash rewards. Visiting Western Defence salesmen like me could quickly make exciting new friends.

Arms Cash allegations in India also swirled around the Indian Barak Missile deal with Israel, which was exposed in a 2001 Media sting operation known as the *Tehelka Tapes*. Mr R. K. Jain, Treasurer of the Samata Party, part of the ruling Coalition, was recorded by Reporters boasting that he had acted as an Arms Cash Agent in the Barak deal, and that he'd secured 3% of the contract value for Minister of Defence George Fernandes and Jaya Jaitly, President of the Samata Party, while keeping 0.5% personally. Many senior Politicians were forced to step down, but incredibly, the investigation was dropped after seven years with no convictions owing to 'lack of evidence'. Nothing to see here.

With around 1.4m personnel in their Armed Forces and the world's third largest Defence budget at over $70bn, everything about the Indian Defence market is big, and the former Jewel in the Crown of the British Empire also has its own network of Ordnance Factories (IOFs) based upon the original British model. I once hosted a UK visit by the Chairman of the IOFs, but in 2009/10 a subsequent Director General was accused of taking bribes in a major Arms Cash investigation and several international Arms Companies were blacklisted as a result.

The gradual release of central Government control over Indian Defence production is stimulating a fast-growing and dynamic hi-tech private Defence sector, as India strives to be self-sufficient in Defence Technologies, so Western Defence and Aerospace Companies really have to invest locally to make any progress; simply turning-up and expecting to sell stuff just won't work, a more nuanced approach is essential. Niche gaps in indigenous capabilities leave some opportunities to be filled by Defence Imports, but India is a totally different Defence sales proposition to, say, the Middle East.

My first scheduled meeting was with resident corporate marketeer and local linkman Roddy Duncan, who was in-country to guide me safely through the complex Indian Procurement waters. After enjoying an Indian breakfast and another slow read of The Times of India Arms Cash article, I waited in the lobby for the arrival of this corporate marketing God, and when he finally appeared (late) there was no mistaking him. Duncan entered through the Hotel's open front door, returning the doorman's immaculate salute with a loose upward flick of his hand by way of response, much as he might have acknowledged some Lance Corporal's salute during his own previous service as a British Army Regimental Infantry Officer.

Those once famous British 'County Infantry Regiments', for so long the backbone of the British Army, have mostly disappeared now, with many being absorbed into a single regimental entity called 'The Rifles'. Popular rumour sustains the story that this concept was inspired in some vain politician's empty head by the TV series 'Sharpe', which featured the Napoleonic era adventures of 'The Rifles' and be that correct or not, I absolutely buy into the spirit of the story. In a few short months, hundreds of years of British military tradition and distinctive uniforms evaporated, and all the subsequent waffle about

103

the so-called 'golden thread' which links them still, remains unconvincing, especially to the Veterans who lost their Regiments.

Roddy Duncan was certainly distinctive, he was a fashion apparition, an affront to the sartorial senses. He strode across the Lobby wearing a tightly-fitting locally tailored grey 'sateen' Safari suit with the flappiest of lapels, with Sixties style flared trousers to match. Because the suit was so tight, the trousers were a little too short, exposing vibrantly red socks with their American style sock-suspenders which became visible when he sat down.

Unlike the fine wine which Duncan loved so much, his Safari suit had not matured well with age, and his lengthy 'handlebar' moustache would surely have been a winner in any moustache-growing competition. It was a revolting archetype of Raj Regimental facial hair, sporting as it did the remnants of his recent breakfast by way of embellishment.

We exchanged pleasantries and sat at a quiet table to take tea and to discuss the visit plan, but Duncan was firstly and foremostly interested in which Regiment(s) I'd served during my own Military service, and this highlights another of the pitfalls of employing ex-Military men. Many of them just can't let go their military pasts and forget that they have left the Forces. They regard their Defence Industry jobs simply as extensions of their military lives in a different guise, undercover as it were. Typically, these folks want and need to establish your Rank, your Regiment, and whether you know Johnny Blenkinsop-Tiddlytwat of the Umpteenth Dragoons, or some-such Regiment, and whether or not you have completed more operational tours than them. Only once this dreary ritual has been completed can you get on with the business in hand, for you have now been benchmarked.

In Duncan's case there was no actual business in-hand, but lots of benchmarking. He'd prepared a detailed military style induction programme for me which appeared to exclude any actual Customers, and I quizzed him on this point in case I'd missed something.

"So, we won't be meeting IMOD (Indian Ministry of Defence) people on this trip?" I ventured, slightly perplexed, but not wishing to hurt Duncan's feelings at this early stage in our new relationship.

"Good Lord no..!" exhorted Duncan, as if this should have been obvious "...far too early for that. We need to get you out and around the Bazaars first, see some of the Country and so on; get you bedded in..." Duncan wasn't referring here to 'Arms Bazaars'.

"…you won't get in to see anyone here for a while, especially without pre-submitting your biodata to IMOD. There's plenty of time for that, we'll get on to it next time around, it takes ages for them to process appointments."

This was a less than good moment for me. Yes, it had been very nice to travel First Class to India, and yes, the Oberoi was comfortable, and their Doormen were very smart, and yes, I did want to see a bit of India. But the whole point of this visit, like any other, was to identify contacts and start meeting people to work out what we could really sell or make here, if anything. We knew business wouldn't be quick in India and that it would most likely be a technology deal, rather than product sales that could be achieved, but you have to get started somewhere. Where were the opportunities and who were the touch points? If there was really nothing for us, I needed to pack up and go elsewhere.

It was Duncan's job as a regional marketing 'Asset' to pave the way for me, to illuminate the target. But after an hour or so that target, whatever and wherever it was, remained very well concealed. I complied with his familiarisation schedule to keep the peace and out of curiosity to learn what he had to offer, and I wondered what local 'help' we had on the books here. All that resurgent interest in the Bofors scandal ensured, officially at least, that open talk about Agents in India was taboo; they were forbidden-fruit and dangerous to know, but in reality, they were queuing up covertly and competing to be signed. Duncan emphasised the need to be circumspect, but our very first port of call was of course with an Agent who Duncan had just appointed. After that meeting I knew we wouldn't be troubled by any business from this man as he was palpably useless. I wasn't going to work with him and told Duncan so straight away and he was less than happy. They probably had a 'side-salad' agreement, the whiff of Arms Cash favours was in the air competing with the Jasmine.

The Company was allowed though to employ an ex-Indian Army Officer, as a fully legitimate local Representative/employee not Agent, to oversee our small legacy contract, and a small Regional Office had been rented in the smart Golf Links area of Delhi for this purpose. It was generously established with support Staff, to discharge the administrative requirements of the old contract which was in its post-delivery death throes. 'Our' ex-Indian Army Colonel who headed this office was, like most Indian Army Officers, very proper and very straight in everything he said and did, and in a neat coincidence, I learned that the Colonel had visited the Indian Navy's ex-Flagship *INS Mysore* during his Army Staff Course. The ageing *INS Mysore* (now scrapped) was previously the British Colony Class Cruiser *HMS Nigeria*, whose crew during the bitter WW II Arctic Convoy Campaign included my Father. *HMS Nigeria* had been sold to India after the War and enjoyed a second and third combat career with the Indian Navy during the Indo-Pakistan conflicts. Her post-War sale to India was a bigger Defence sales success than I was likely to achieve on this trip.

Languid easy days passed in the Oberoi, where I swam, and circuit trained. I was ready for the Olympics. Outside the hotel I visited places of interest in New Delhi, with my Minder Duncan providing a full tourist running-commentary. He boasted to me about attending the Indian Army Staff College during his British military Service, which explained his part-Raj, part 'gone-native' demeanour. From the back of nausea-inducing Austin Ambassador taxis, Duncan pointed out to me the very impressive and huge Central Secretariat Buildings at Raisina Hill, designed by Architect Herbert Baker, and completed in 1931, apparently in the 'Indo-Saracenic Revival architectural' style and known more simply to everyone as 'North Block' and 'South Block'.

These identical buildings are huge, with around a thousand rooms in each; the Indian Ministry of Defence was in South Block. I also visited 'India Gate', that huge Lutyens memorial to death, which marks Indian losses in several Wars. Indian troops were despatched to Europe to fight for Britain in Flanders Fields and on the killing grounds of Gallipoli, which must have been a totally alien environment to the men who died there in droves; 84,000 Indian troops died in the Great War, and like British Conscripts they did not make sacrifices for Britain, they were sacrificed by Britain.

I could only imagine what glittering Defence business opportunities might be hiding behind South Block's impressive but closed doors, if they could only be prised open. Duncan needed to get his act together and get me inside, assuming of course that we survived the swirling, anarchic Delhi traffic and the heat. But Duncan either could not or would not get meetings with IMOD and advised me yet again that we had to be patient.

The novelty of touring Delhi was waning rapidly, but as luck would have it, Duncan was unexpectedly called away from India to fend off some regional corporate marketing disaster in Kuala Lumpur, so he left me to my own devices. There may be a God after all I reflected, or indeed a selection of Gods from which to choose in India. Duncan's enforced departure coincided with a weekend, so I took the opportunity of Minder-free time to head out to Agra and visit Shah Jehan's masterpiece, the Taj Mahal. We were scheduled to go Agra anyway later during my visit but getting there without Duncan was preferable.

The drive to Agra was everything I'd imagined a long drive in India would be; frequently dangerous but mostly just unpleasant, except for occasional visual landscape treats. The broken-edged crumbling roads were patrolled by Camels and festooned with primitive carts of all shapes and sizes, overburdened with cargoes of every description. Ancient Pick-up Trucks chugged along on axles bending beneath huge loads, spewing forth clouds of toxic smoke, whilst working elephants lumbered dismissively into the paths of oncoming cars. The whole scene was crisscrossed by a guerrilla army of feral scavenging dogs, whilst skinny cattle wandered unhindered wherever they pleased. Inevitably, there were numerous traffic accidents.

Improvised shanty buildings had sprung up everywhere, constructed from scavenged scraps of wood and random pieces of junk; nothing went to waste here except humanity. Enterprising souls retailed soft drinks from ancient grubby-looking cold-boxes at the roadside, and communal outdoor TV screens attracted crowds to watch the latest episode of some Bolly-Soap opera or other; escapism for poor people, and there were plenty of them, and much from which to escape. What lodged and remained in my mind's eye above all other things, was the acrid stench of unrelenting, unmitigated poverty. I'm sure the Indian National Space and Nuclear Weapons programmes were sources of great comfort to these folks and their skinny barefoot kids who ran around aimlessly; they certainly weren't pursuing any aspirations.

By way of contrast, rich was the experience of visiting the Taj Mahal itself. On arrival I was besieged by a stampede of Guides each offering me a unique guided tour at some bargain price. They pressed in tightly from every angle, until I selected one old man as my Guide, and we went off together into the curtilage of the great monument. I didn't know that Shah Jehan had planned and started a symmetrically matching building for himself on the opposing banks of the Junuma River, but this hadn't been completed when he died. He must have used British Building Contractors.

My Guide had escorted visitors around the site for thirty-eight years and he'd perfected a gentle, subtle manner which I rather enjoyed. He paused occasionally to point out some super viewing angle, or to suggest a good place for a photograph. He framed views for me beneath pointed arches and so on, but I guess all Guides there do similarly. It's not within me to 'gush' over artefacts, buildings or views, but I had to acknowledge the Taj Mahal was one of those rare buildings and locations where reality exceeded my expectation. Sadly, the great monument was degrading at the time of my visit owing to the discolouration of its decorative tiles by emissions from a nearby Chemical Plant. I suspect Shah Jehan would have fixed that quite quickly. The return journey to Delhi was hell on wheels in the dark, much as anticipated, so the charm of the earlier day evaporated quickly.

Once back in the Oberoi, the preparation of my Indian Defence Business action-plan was put on hold by a more immediate, more pressing personal problem. I'd been circuit training in the hotel gym and finished my evening session with a swim, and once back in my room I took a pre-dinner Shower. I was holding my face up into the flowing water, when a horrible rancid stench suddenly rose from the water and engulfed me. Instinctively I pulled my face out of the flow, spat, and opened my eyes. The shower water had turned murky brown, almost orange in colour, and smelled of sewage. I jumped out of the shower cubicle shaking my head, and trying to throw off any remaining water from my skin, and I rinsed myself quickly with chilled bottled water from the room fridge, all that was available, and called Reception for more clean water and told them about the shower.

In the early hours of the morning everything went wrong. I wasn't simply unwell, that would be a major understatement, and such was the acute pain that I really thought I was dying; it was a completely debilitating situation which lasted for several days of sweating and shivering as I drifted in and out of reality in a state of delirium. A Doctor attended frequently. This was more than good old 'Delhi-belly' since it transpired that seepage from local landfill sites had infiltrated and contaminated the water supply, and some very unpleasant 'matter' resided in these landfills. Anything that moved and died but couldn't be eaten went into landfill, along with all the usual other debris. The Doctor eventually got things under control, and having endured a severely weakening bout of illness, I promised myself that I would get in front of an Indian Customer before India killed me, then get out of the damned country as quickly as possible.

A week later I was able to visit the British Defence Attaché and the First Secretary, Defence Supply (FSDS) for a chat, I needed to get their take on the Indian Defence business situation and find out how to go about getting some real IMoD contacts, if indeed that was at all possible. The FSDS was laconic and realistic, or should that have been fatalistic?

"I don't know what you think you're going to sell here..." he started off discouragingly "...they have thirty-nine Ordnance Factories and manufacture everything from boots and ballistic missiles to nuclear weapons... if they can't make it themselves, they get it from Russia, or anywhere else where it's cheap and available..." he added.

"...occasionally they'll buy from the West, but only for things like Mirage jets or Submarines, just the things they can't produce yet. Be ready to invest and Team-up here with local partners or fail. Talk to Bharat, talk to Tata etc."

Like many visiting Defence Executives before me, I hadn't fully appreciated from afar the extent of the dynamic growth in indigenous Indian capabilities. Long gone were the days when India had to purchase equipment from the West because they had no choice; they were established and successful Defence Exporters now, and competitors in some product areas. UK MoD had even bought Small Arms ammunition from India and I'd used it during my Army service, it was awful. I re-calibrated my expectations as any significant progress would have to be for something really special, something high-tech and long-term with a local Partner Company. Roddy Duncan had missed the key point completely; we weren't going to sell products here anymore.

The First Secretary gave me his Guide to Defence Business, which contained useful looking contact numbers and I went back to the Oberoi to read it from cover to cover, which didn't take long. I started calling the numbers. The First Secretary's sage advice rang in my ear '...remember, getting a signature on a Contract here in India is only the first stage in negotiation...' I rang someone in IMOD, as described in the Guide, and arranged an appointment straight away.

"Don't you need my Bio-Data?" I asked the Official, conscious of Duncan's remarks to me at the start of my visit, when he said this was mandatory.

"No old chap..." came the cheery reply "... just come along and bring your Passport. We'll be pleased to see you."

So along I went, Passport in hand, and walked straight into IMOD where I met a senior Official with responsibilities for foreign Defence Imports. He was helpful and good humoured, and he fixed other appointments for me, including with the Army Engineers (strange dark meeting room, all rather spooky, probably recorded) so at least I'd met a potential Customer. It might have been chickenfeed in the wider business context, but it was a start. I could see more clearly than ever that only high-tech stuff would be of the slightest interest here and getting Export Clearances for that at home could prove challenging. These days the MoD/UK Government happily donates millions of Taxpayers' pounds worth of munitions to kill Russians, but other things need consent.

When Roddy Duncan finally returned from his Malaysian troubleshooting visit, he was furious about what he regarded as my maverick activity. Visiting the Customer like that was just not how things were done here in India, he told me forcibly. I told him to f*** off. It was a common spat 'twixt an operational Division with targets to hit, and a central Corporate Marketing team which lived in dreamworld. The issues between us were about Company protocols not personalities, although in Duncan's case it did come pretty close. I've encountered similar situations in almost every major Defence Company I've worked for, and that's why working Freelance is such a joy, you don't have the same limitations, and if they do pop up you just drive over them. My first visit to India had been an eye-opener in many senses, but before leaving, I couldn't resist a parting-shot at Duncan, the 19th Century Safari-suited snob.

Duncan asked me what else I'd got up to whilst he'd been away pouring oil, or maybe petrol, on troubled waters in Kuala Lumpur, so I told him that I'd been invited to drinks at the Gymkhana Club, one of the oldest and socially still most prestigious Clubs in India. The Gymkhana Club was at the very top of Duncan's desirable list of places to go in India, but sadly for him he'd never been invited. During my First-Class flight to New Delhi, one of the two other people with whom I'd shared the Cabin had been a young Indian Businessman, self-described as being of high Caste.

We chatted a lot about business and politics during the long flight and we got on very well. Before we parted on Arrival at Indira Ghandi airport, he invited me to the Delhi Gymkhana Club (in Colonial times called 'The Imperial Gymkhana Club') as his personal Guest. Duncan was fuming because I wouldn't tell him who'd invited me, and the very idea that I'd been to the Club at all caused him anguish. It was, I kept telling him, purely a personal invitation between friends; nothing to do with him or the Company. He'd probably have tortured me to find out who my host had been, and I delighted in his angst; petty, I admit, but fun at the time.

From a purely business perspective, my first visit to India had been one of the most frustratingly useless and expensive of my Career, although from a personal point of view and illness excluded, it had been a fascinating first visit to the sub-continent. However, I was in no hurry to go back. Inevitably the infamous Bofors scandal was joined by similar corruption cases in India, including the so-called 'Choppergate case' which came to light in early 2013 when a parliamentary investigation opened into allegations of bribery and corruption involving helicopter manufacturer Agusta Westland and various senior Indian officials.

The investigations related to the purchase of a new helicopter fleet, and Indian Defence Minister A.K. Antony confirmed the corruption allegations, commenting on 25th March 2013:

'Yes, corruption has taken place in the helicopter deal and bribes have been taken. The CBI is pursuing the case very vigorously'.

Italian police later arrested Finmeccanica's CEO, and the Italian Prime Minister Mario Monti, stated that

'There is a problem with the governance of Finmeccanica at the moment and we will face up to it.'

So perceptive.

The later Indian purchase of French Rafale jets was similarly mired in controversy with allegations of corruption and French media website Mediapart reported:

'The Narendra Modi-led government is refusing to cooperate with French judges who have requested India's assistance in their ongoing investigation into alleged corruption in the sale of 36 Dassault-built Rafale fighter jets to India in 2016 for €7.8 billion...'

Different Military Jets were linked to corruption allegations in May 2023, when Al Jazeera reported that:

'India has filed a graft case against Britain's BAE Systems plc and Rolls-Royce Holdings for criminal conspiracy in the procurement and licensed manufacturing of 123 advanced jet trainers, according to a Federal Police document. The case is based on the findings of an investigation launched by India's Central Bureau of Investigation (CBI) in 2016...'

Al Jazeera further reported that:

'In 2012, Britain's Serious Fraud Office (SFO) launched an investigation into Rolls-Royce for payments linked to transactions with countries including China, India, Indonesia, Malaysia and Thailand. Rolls-Royce paid a fine of 497 million pounds ($614.19m) to settle that case with the UK SFO in 2017.'

I didn't get back to India for some time after that first visit, and when I did, I went to Mumbai for an altogether different purpose. In that City in which the richest-of-the-rich live cheek by jowl with the poorest-of-the-poor, how the poor could do with even a fraction of the wealth exchanged as 'Arms Cash'. But I was headed South next, and my new assignment would carry me beyond India, and across the Gulf of Mannar, to Sri Lanka.

Postcard from Colombo
I flew into Colombo from Dubai for meetings with the Sri Lankan Army, and in some respects, it was like my first Business trip to India. On that occasion I'd been one of just three First-Class compartment passengers, so I'd been thoroughly spoiled by the Cabin Attendants, and this time, I was the sole Club-Class passenger on a new Airbus, so I was thoroughly spoiled once again. The slightly diverted flight path routed across southern Iran and skimmed the edge of the Silk Road, before arcing south across the Arabian Sea to the island paradise of Sri Lanka, formerly called Ceylon, and reputedly the original Garden of Eden. Adam and Eve may have had apple trouble, but they weren't caught up in a vicious Civil War as Sri Lanka was then, and I was surprised that lethal Defence exports from the UK were still being authorised. I'd done my usual pre-visit research, and read all the Company files on Sri Lanka, as well as getting a Brief from the Government's Defence Export Services Organisation (DESO) Desk Officer.

DESO was a controversial Government Department from the 'get-go' in 1965, when it was established under Harold Wilson's Labour Government. Ironically, it was Gordon Brown's Labour Government which announced DESO's closure in 2007, with many of DESO's functions moving to UK Trade & Industry Ministry (UKTI). Today 'UK Defence & Security Exports' (UKDSE) is part of the Department for Business and Trade.

In 1965 there had been supra-excitement, or more probably economic desperation, over the Export revenue potential of British Arms Sales, and Lord Stokes was tasked to investigate the subject. His subsequent Report noted that Commissions, i.e. Arms Cash, were frequently paid during Arms deals, and that in some cases Defence equipment was ordered solely so that Commissions could be paid. I certainly recognise that scenario.

DESO consisted of London-based Civil Servants organised in regional desk-teams, linked to in-country UK Defence Attaches. Although it was DESO's job to advise and support exporting British Companies on Defence business opportunities overseas, there was no requirement for DESO Staff to have had any actual personal experience, either of the regions they covered, or of the Defence Industry generally, or even simply of Business. Unsurprisingly therefore, most DESO Civil Servants I met were completely ill-equipped for their roles. It's a massive irony today, that 'Stormin' Starmer's Labour Government has rediscovered an appetite for Economic growth through Arms exporting, which Harold Wilson championed. Today, the Defence Industry is described by the Government as 'an engine for economic growth'. What a morally depressing statement that is. At the most senior level there was 'revolving-door' activity between DESO and the Defence Industry, which raised serious conflicts of interest issues, as Senior DESO Officials pandered to their favourite Defence Contractors in the hope of being offered well-paid private sector jobs.

I duly reported for my Sri Lanka briefing in Soho Square, DESO's old HQ location before they moved to Victoria. I preferred to keep DESO at arm's length because it leaked like a sieve, with Officials seemingly passing on whatever they gleaned from one Company to another. They liked to gossip. Well, let's face it, they had nothing better to do. A spotty youth, who'd never been to Sri Lanka, delivered the Country Briefing to me, although I suspected that he'd have difficulty in locating the troubled Island on a map. He waffled on for about forty-five minutes saying little of interest, and when he finished, I asked him about the Defence Attaché in Colombo.

He was very disparaging about the Officer in question, and the ex-military part of me winced as it wasn't really this Boy-Wonder's place to make those comments. Any serving Officer of Colonel's rank would have considerable operational experience, whilst the DESO lad probably seldom ventured past Waterloo Station on his Scooter. He was arrogant and unpleasant, so I made a mental note to 'drop him' when the opportunity arose, and if a moment didn't arise, I would create one.

The biggest surprise was that I was allowed and even encouraged to go to Colombo in the first place, as the Sri Lankan Forces were in the midst of operations against the Liberation Tamil Tigers of Eelam, the 'LTTE', who were waging a determined and bloody campaign against the Government. It was a full-on Civil War and I was marketing lethal products. I thought UK Policy was to desist in such situations, but as we all know these days, the UK Government delights in escalating and prolonging other people's wars and increasing casualty lists by supplying additional weapons and equipment, as in Ukraine and Israel. UK Taxpayers must be proud of their Government's role in killing Russians. Was that in any Political Manifesto?

The Airbus wheeled and descended gently in the clear morning light, and during the final approach I had a very good view of everything below; it was an impressively soft touch-down, which frankly was a relief in view of the recent Airbus crash in India. The first thing on the ground which caught my eye was a flight-line of ex US Army Huey helicopters, which with their jungle backdrop, could have passed for a scene from Vietnam. I'd enjoyed flying in Hueys during my military service in Oman as they felt nice and stable and were easy to rope-down from. I zipped through Customs and Immigration with no undue delay or interrogation and hopped into the back of an old black 'Morris 1000' taxi for the bumpy ride to my Colombo hotel.

Two things sprang to mind as my taxi journey progressed. Firstly, the whole place had that casual, unkempt gradually-falling-apart ambience about it, which became apparent even in the short distance I'd travelled from the airport. The second, was how perfect the road was for ambush purposes, a gift in fact. With my backpack stuffed full of Defence equipment information, it wouldn't be a great time to meet the LTTE, who were busy fighting a desperate but successful guerrilla campaign. Militarily theirs was the side to be on, just as former US Green Beret Officer Larry Lanning once observed in an article about the Vietcong, which I read when I was a teenager. As an adult I read his excellent book about Military Commanders.

During my army service in Oman, my Sri Lankan Clerk had told me all about the LTTE, in fact he evinced considerable respect for them. They were, he told me, doughty fighters, highly determined and ideologically motivated. They were ready for, and often committed suicide attacks before the practice was adopted by Islamic Radicals. Many of the 'Tigers' carried cyanide capsules, preferring death over capture and interrogation, which Israeli Advisers were assisting with apparently. An ex-Oman Army friend of mine who served in Sri Lanka after his Oman Contract, found duty there to be too far outside the boundaries of the Geneva Convention for his taste.

I arrived unscathed in the centre of Colombo and checked-in to my Hotel, which maintained a Security Floor for Western visitors like me. Oddly, there were very few other tall, white, western businessmen in lightweight business suits strutting around Colombo, so I stood out like dog's balls. I dressed down accordingly, striving in vain for the 'grey-man' look so beloved of Hereford. It didn't work of course, I still stood out like dog's balls. I did my standard recce of the hotel, checking entry points, exits etc., and made a room blocking plan. A new Tamil phenomenon had gripped the Capital at that time, the so-called 'Suitcase Sniper'. These terrorists re-employed the old Chicago Gangster movie tactic of carrying a bag or package (Violin case? lol) of some sort to conceal a weapon, and loitered on street corners waiting for predicted or for opportunity targets. It was hardly sniping in the modern refined Western military sense, but it worked well enough for them.

This all concentrated my mind wonderfully on personal security issues, especially since I was due to visit the Army Headquarters in person to discuss and promote my (lethal) products. Had I been a Tamil Tiger, I reckon I would have maintained surveillance on the approaches, entrances, and exits to and from the Ministry of Defence buildings and Army Offices, to see who was coming and going and to follow up with a shoot. I didn't know for sure if they did or didn't do this, I simply assumed it happened; I would have done it if I was a 'Tiger'. There were only a few hotels in Colombo in which Westerners stayed, so it would have been easy enough to follow anyone for assassination purposes. It was still hard to believe that the UK Government sanctioned the export of lethal products to this Country, but my job was to simply sell more, and I was very good at my job.

I made several visits to the Headquarters and varied my routine as much as possible in micro and macro ways. I took a cab from the Hotel, I took a cab from a different Hotel, I dropped off short and walked in, I drove in the whole way. I wore a suit or just shirtsleeves with slacks, sometimes glasses, no glasses, a bag no bag, a hat or no hat, I waved to imaginary people, I used a direct route back to my hotel, an indirect route, I went to a different hotel first. I made morning and afternoon appointments. I tried to improvise as much as possible using any and every little variation of routine and appearance which might inhibit or confuse surveillance for even a moment; a fleeting second may be enough sometimes to create that critical hesitation which inhibits a plan. All this didn't amount to much of course, but remember, having limited options does not mean 'do nothing' and after several days of this it was pleasant to take time off.

I'd been travelling for several weeks and had spent far too much time in hotels or on planes or in Offices, so I was ready for some exercise and the hotel had a very well-equipped Gym. After a leisurely breakfast I grabbed my training kit and headed off to the gym, which was empty, much like the rest of the hotel, and I warmed up and stretched, then launched into my circuits. The travel-induced rust fell away grudgingly, but I had the place to myself and that was a joy; no time lost to Lycra-clad body-building narcissists hogging the equipment I wanted to use, and none of that fashionable but comical posing in mirrors.

I was in mid-Circuit when I became conscious of a second person entering the gym, and I caught a sideways glimpse of a tall slim young woman stretching. She started her circuit and drove hard at it, tearing into explosive weight training repetitions, and beating the shit out of the punch bag with Martial Arts kicks and punches. A competitive atmosphere settled over the Gym as we pushed weights and forced out

116

extra sit-ups; no prissy spinning or treadmills thank you, just actual training.

In a brief sweaty coincidental moment, we paused, gasping for air and dripping sweat and exchanged poker-player glances. After some verbal fencing she told me she worked in the Israel Bureau, and the penny dropped. Although there'd been some diplomatic friction between Sri Lanka and Israel, relationships improved to the point where an Israeli Interests office had opened within the US Embassy in Colombo, which is presumably what my new training companion meant when she referred to the 'Israel Bureau'.

Stories were circulating about Israeli personnel interrogating LTTE prisoners and Israeli-supplied Drones played a critical role in the war. The Sri Lanka Government also bought Israeli Kfir jets, plus naval vessels and Artillery systems, and Colombo was awash with all sorts of Arms Cash rumors. Other rumors concerned the re-cycling of weapons from Asian countries to the LTTE, and of course it was suggested that Mossad was linked to that too, just as Mossad is casually and imaginatively linked speculatively to a whole raft of incidents and events all around the world, whether true or false. Who knows what really happened in Sri Lanka, I certainly didn't. I could only speculate as to the likely role of my feisty training companion. We had dinner together. The next day I was due to meet the UK Defence Attaché.

At the Attaché's suggestion we met at his Residence, rather than in his Office, and I enjoyed a Raj-style afternoon Tea on the Veranda with him; I temporarily put out of my mind the disparaging remarks made by the London DESO Briefer. I met a lot of DAs, so I could make comparisons, but I wasn't there to judge or rate anyone, I just wanted information to help me target and win Defence Business. What I got from the Colonel was the single most comprehensive and insightful Country briefing I've ever had, or subsequently have ever received from any Defence or Military Attaché anywhere.

He knew who-was-who in the various Departments and sketched out the likely political succession also; it's important to know who's likely to be in charge tomorrow. He also knew who exerted real influence and who simply postured and so on. It was priceless as far as I was concerned, and I also enjoyed the Tea. The DA was a convivial host and was probably pleased to meet any occasional British visitor and I mentally reaffirmed my other mission and repay the DESO boy on my return.

It was a darkening but pleasant early evening when I left the Attaché's Residence, so pleasant in fact that I opted for a short walk, though not the whole way back to the hotel some way distant. I'd walked barely a kilometre when I heard it. That distinctive sound was known to me; it had been filed away in my acoustic archive since the first time I'd heard it years before in Imber Village on Salisbury Plain.

Imber is the little Wiltshire village, which was requisitioned, most would say stolen from its rightful inhabitants by the British Government, as a temporary military training area during World War II. It was never returned to the Community and continued to be used for Army training. Many years after the World War ended, training for Northern Ireland operations was conducted there, and when that conflict intensified, additional ugly concrete-block buildings sprang up to supplement Imber's original traditional village structure.

Conventional 'Fighting-in-Built-Up Areas' training in preparation for Soviet aggression, and later, for the Balkans conflict, was also carried out in Imber, until the Army's purpose-built Facility at nearby Copehill Down became available, many years too late of course for its original design purpose. Sad little Imber continues to this day to be used as an Army Training location, over eighty years after it was requisitioned with a promise to return it to the residents six months after it was taken from them. Only the little village Church of St Giles remains wired off and unmolested by the military. It stands alone and sad, off-limits to marauding troops, although it does get hit occasionally during training.

It was in Imber Village during an Infantry Tactics Course that I was first exposed to the firing signatures of a variety of Small Arms weapons. Selected weapons were fired low overhead, in close proximity to captive, and very attentive audiences of young Infantry Officers. The IRA ensured that many of us soon became depressingly familiar with the American M16's high-velocity whip-like crack, whilst the Soviet AK 47's distinctive signature could be heard in most conflicts around the world. But the sub-sonic bang and slow-moving big and meaty 'slug' of the .45 calibre Thompson machine gun, was something distinctive. A sub-sonic round has no whip-like crack overhead, but when the Thompson was fired through the undergrowth

118

during the demonstration, the young Infantry audience could almost feel the heavy, slow-moving projectile tearing and shredding foliage, and smashing through light branches as as it would smash into flesh and bone. It was acoustically memorable, and I filed it away hoping never to hear it again.

On that darkening Sri Lankan evening I did hear it again, the unmistakable signature of a 'fat' low-velocity round being fired and tearing its way through dense foliage. Luckily for me, the Tactics Course had been a Summer one, otherwise there would have been no foliage. Several more rounds were fired, singly not in bursts, and instinctively I dropped down beside a convenient shadow-casting wall. These rounds were not being fired at me. There was sporadic return fire, a few more low-velocity rounds fired from a single shot weapon, a pistol probably. Then it all went quiet again.

The next thing I heard was a Trishaw approaching; one of those gently motorised chugging tricycles with colourful canopies, which serve as cheap and cheerful Taxis in Colombo. In a moment of pure comedy, I jumped aboard this Trishaw and ordered the driver to make best speed for my hotel. It was slow, and I could probably have run faster, but it served its purpose. I cleared the site of the 'contact' to return intact to the comfort of the hotel's Security Floor.

I saw no merit in becoming the meat in someone else's ambush sandwich in the struggle 'twixt the Tigers and Sri Lankan Government Forces, so having finished my business, and having re-focussed my local Agent, who seemed to be earning more from marble and timber exports than from Defence equipment, it was time to leave. This conflict was not being settled according to Queensbury Rules, well, which Conflicts ever are? In the historic words of Admiral of the Fleet John 'Jacky' Fisher, who dragged the Royal Navy kicking and screaming into a new era before WWI, 'The Essence of War is Violence. Moderation in War is Imbecility'.

I still couldn't really understand why UK Export Licences were still being granted to Sri Lanka for lethal products during a Civil War, or indeed afterwards, despite the Sri Lankan Government's appalling human rights record. In 2013 alone, two million pounds worth of security exports classified as 'lethal' were exported from Britain to Sri Lanka, and even another article in The Guardian could not arouse enough indignation to provoke feeble Politicians to address the situation.

Sri Lanka was a small example of this issue relatively speaking, but clearly, no one in Government circles gives a fig about this stuff, and probably no one in Government really understands it either. There's no point in living in hope of change though, since 'hope' is not a strategy. The supply of lethal products to export Customers from Britain has become many times worse since my distant Sri Lanka visit days, and not simply because of the Ukraine and Gaza situations, although they represent new highs. British Taxpayer funding enabling the killing people in other Nations has now risen into billions of pounds worth of arms, ammunition and military materiel, with less checks and balances than ever before. Arms Cash issues seem insignificant by way of comparison with the new Political appetite for sponsoring deaths in other States. Once weapons and munitions are delivered, the Supplying country has no control whatsoever of the use thereof.

I was happy enough to leave the troubled Island Paradise which my Father had visited long before me during the second stage of his World War II service, this time aboard the Aircraft Carrier *HMS Unicorn*, which he'd joined following his previous service in the Arctic Campaign. *HMS Unicorn* had neither been designed nor equipped to service and support the huge variety of British and American aircraft types she found herself supporting in the Pacific campaign, but the ingenuity of her Fleet Air Arm crew made most things possible and kept aircraft flying in arduous circumstances.

Judging from the performance of the brand-new Aircraft Carrier *HMS Prince of Wales* it's now the whole Carrier which breaks down these days. This taxpayer-funded, technologically advanced (?) and massively over-budget Aircraft Carrier, which apparently cost Taxpayers around £3bn, barely cleared Portsmouth harbour on her maiden voyage to the USA, when she broke down and had to be towed away for nine months of repairs. The rectification work alone cost around £25 million, an amount which was almost ten times what it cost to build the WWII HMS *Unicorn*, which managed to steam over 400,000 wartime sea miles without a hitch, and was still ready to go to War again in Korea years later. There must be a lesson there somewhere. I settled back alone in Club-Class comfort once again and reflected on the technological gulfs between the WWII aircraft carried aboard *HMS Unicorn,* and the 'fly-by-wire' digitised Airbus in which I now flew. But it was aircraft of a different type which had set the Arms Cash tills a-ringing back in Colombo.

The acquisition of Ukrainian MIG military aircraft by the Sri Lankan government attracted intense media scrutiny in the Colombo Press, because standard Government Defence Tender processes were mysteriously side-lined. The Sri Lankan Air Force wanted additional planes to bomb the Tamil Tigers, and the MIGs were to be supplied originally to the Sri Lankan Air Force by the Ukrainian Government Agency UKRINMASH.

A subsequent international investigation revealed that the Agency had sold the aircraft instead to a Singapore-based Company called D.S. Alliance, instead of directly to the Sri Lankan Government. In this deal, the same aircraft as had been originally scheduled for direct sale to Sri Lanka, were sold indirectly at a very a high price for ageing airframes, in what became a 'Government-to-Government Agency-to-Company-to-Government' deal.

The price paid under these arrangements was significantly higher than had been paid directly by the Sri Lankan Government for similar aircraft some years earlier. What added the spice of an Arms Cash mystery to this deal, was the discovery that payments in the deal were being routed through a mysterious Bank account in London in the name of Bellimissa Holdings Limited, which turned out to be a shell Company based in the British Virgin Islands. Why would any Sri Lankan Government Defence purchase need to be financially routed through a non-involved third-party Company in a foreign Country? Take a guess.

The payments were to be made on a two-year schedule when five years was anticipated, and it reminded me of an Indonesian deal I experienced (recounted later) insofar as it had all the hallmarks of a classic Arms Cash 'cut-out' and it fitted the profile for an Arms Cash 'Influencer' at work. Someone was able to make a high-level, expensive and technically inexplicable strategic purchase decision unchallenged, combined with eccentric financial routing. Not so dissimilar from some of the UK's PPE Procurement decisions during the COVID pandemic.

It was discovered that the Singaporean Company owners of D. S. Alliance had purchased the MIGs from UKRINMASH for just over $7m, then sold them on to the Sri Lankan Government for over $14m. Nice margin for old planes. When Sri Lankan Detectives started to follow the money trail, their enquiries led them eventually to an Apartment in Dubai, where local Police arrested Sri Lankan middleman Udayanga Weeratunga, but further Investigations were closed down. Media probes came to an abrupt halt when Lasantha Wickrematung, the Editor of the Sri Lankan newspaper 'The Sunday Leader' who was investigating the story, was murdered. His killers have never been brought to justice. Like my own visit to Colombo, the matter came to an inconclusive end, and I left Colombo with my primary memory being that of an energetic and feisty training companion from the Israel Bureau. I headed off, completely coincidentally of course, to Tel Aviv.

Chapter 5
Around the World in eighty deals

It's hard to imagine that any other Industry would so willingly have cut me loose overseas so widely and so frequently on business missions. I've travelled to around forty countries in pursuit of Defence, Security and Military Aerospace business, whilst working for major and minor British, French, American and Canadian Companies, either as an employee or as a freelance Contractor.

International travel was always magnetic for me, and since I first plunged into the murky waters of the Defence world, some Countries once deemed acceptable as great Customers, became pariahs overnight, Iran for example, and changes to legislation undoubtedly impacted on potential Arms Cash arrangements, though not always as Legislators had intended. In some Countries where Arms Cash had previously been part of the fabric of normal business, Defence Procurement postures changed with the adoption of more transparent processes and tougher legislation, whilst in others, it remained Arms Cash Business as usual.

It's not always essential to actually visit a country in order to sell there, as will be evident from some of the following Postcards, but international travel on Defence Business can offer an exciting cocktail of risk and opportunity, with the potential for exotic personal encounters, not all of which result in death threats; on the other hand, some do. If you don't have a highly developed and dark sense of humour this is not the Industry for you.

British-based Defence Companies traditionally followed the 'Empire flight path' diagonally across the Globe to the Middle East and thence to India and on to Singapore and the Antipodes, and many home-based Senior Executives seemed impervious, and still do, to the changing geo-political scenarios which rendered many of their missions either improbable, or sometimes impossible, even before they started out.

Perhaps, like me, they were happy to travel simply for the love of travelling and the stimulation of risk and exploration, unlikely of course, or more simply to satisfy an unquenchable thirst for Air Miles or Arms Cash backhanders. Why else would anyone with half the brain of a tomato want to do this work, unless possessed of a curious psychotic appetite for stress and drama? For those in need of both sensations, where better to go than Tel Aviv?

Postcards from Tel Aviv, Jaffa & Jerusalem (pre-Oct 7th - Gaza)
My first Defence Business visit to Israel was triggered by an informal meeting at a major Defence & Aerospace Show, and until then, Israel had been off-limits for many of the UK-based Defence Companies I'd represented. I'd been chatting with an Arab Delegation on a Corporate Aerospace Exhibition Stand when I noticed a man bobbing around behind some other Visitors and trying to catch my eye. I slipped away to find him, but he found me first and pressed his Business Card into my hand. Abel from Tel Aviv. The address made me smile. When he explained who'd given him my name, I knew it was OK, or rather that *he* was OK.

Away from the Stand we sipped truly awful Exhibition coffee and chatted about Military aerospace stuff. Abel wanted to make a Proposal for an upcoming Israeli Air Force Project and although his Company in Tel Aviv was already doing business with the IAF, he needed a bigger Partner Company to team up with for this new Project. Such is the reputation of the IAF that approaching them with a Military Aerospace Training Services Offer, seemed to be a bit like a carrying-coals-to-Newcastle challenge, but Abel explained why it was a good idea and there was a definite hook there, something to go for.

You don't always have to break down doors or sign-up Arms Cash Agents to get 'in' with a Defence Customer, and Teaming openly and legitimately with a local Company in Israel, would be a quicker, smoother approach than any I could make directly. The 'internal sell' within my own Company would be the harder part. Abel and I matrixed our respective Company strengths and I had to consider carefully what would and wouldn't be acceptable to my Company. Most UK-based Defence Companies at that time were very nervous about even travelling to Israel, never mind bidding or contracting there, as the prevalent belief was that doing Defence business in Israel would torpedo their lucrative business in Arab States. Ask them about that again today.

I ran some project concepts past a Main Board Director and was cleared to engage further, which frankly was a shock, so I booked a flight to Paris to meet Abel again, this time at the scorchingly hot Paris Air Show, where we added commercial texture to our Project concepts.

We compensated for suffering in the Air Show heat, by retreating to the centre of Paris to eat extraordinarily well in Abel's preferred seafood restaurant. The Lobster there was unworldly, a spiritual experience of a dish. Food, wine and Abel were constant and reliable companions I'm glad to say, and he was also a pragmatist of a Businessman with robust and healthy Aviation business interests in Tel Aviv. Exactly the sort of individual who scares the pants off most institutionalised Defence and Aerospace Company Plc Commercial Departments. The Company 'Bid Police' weren't going to enjoy this one, but there again, they didn't really enjoy anything, apart from soft, fat UK McD Contracts, and they could even screw-up those when left unsupervised by adults.

Aside from their recurrent fear that business in Israel would scupper their lucrative business with Arab countries, many UK-based Defence Companies believed it to be completely pointless anyway, because they would always lose in competitions on Israeli home turf. There was/is a widespread belief that corruption held sway in the Israeli Defence domain, but Arms corruption allegations in Israel were/are usually strangled at birth by comprehensive legal gagging orders. The respected Israeli Haaretz Newspaper stated that:

'...when it comes to the Israeli Military, Intelligence and Security bodies, the Police, the prosecution and the courts show less determination. They are softer and more flexible - and thus, they hurt what must be an uncompromising battle against corruption - the misuse of power and of public funds.'

British Defence exports to Israel rated at £387m-worth being supplied since 2015, and they have surged recently, but this is small change compared to the huge amounts UK MoD now spends on imported Israeli Defence systems and project participation. At time of writing it is impossible to predict the impacts of the Gaza events on UK-Israel Defence contracting, but the UK Government assured Israel of seemingly unconditional continued support. In October 2023 a spokesperson for the Campaign Against the Arms Trade observed:

'It is disgusting that the Department for Business and Trade is refusing to suspend and review arms licences to Israel given the mounting evidence of war crimes committed by Israel in Gaza.... UK industry is responsible for 15% of the components used in the F35 stealth combat aircraft that are being used in airstrikes, *and the UK is therefore complicit in war crimes committed by the Israeli government.' It seems that the UK Government did not even consider suspending or sanctioning Israel's growing participation in UK Defence Procurement as a tool for applying political pressure either.*

From a professional perspective, I can say that Israeli Defence Companies competing in international markets are notably hungrier, better focused, more assertive, more industrious and less risk-averse than are their UK counterparts, who therefore get what they deserve in encounters with them; i.e. second place. These attributes, coupled with increasingly close political ties between UK and Israel, may explain why Israeli Defence Companies have achieved such rapid and otherwise surprising deep penetration into UK's domestic Defence markets so rapidly in recent years.

UK MoD spent a reported $1.25bn with a Thales (France) & Elbit (Israel) Consortium for the 'Watchkeeper' Drone, which is based on the Israeli 'Hermes' system and which was widely used in Afghanistan by British Forces. Israeli Drones have been used in targeted killings and assassinations, so their sales mantra might well be *'Proven on Palestinians, bought by the British'*. But the UK Foreign Office and MoD are tone deaf to the downstream impact of all this, and the increasing use of Israeli-produced Defence equipment and Systems by British Forces will hardly endear British Service personnel to Muslim Customers, be them either allies or enemies, in future conflicts. Why British Companies, once world aviation pioneers, could not even build a combat Drone remains a major mystery.

UK MoD also bought the Israeli Rafael anti-Drone system in a contract reputedly worth $20m, while in 2019 MoD also signed up a $38 million contract with Elbit for a Joint Fires Synthetic Training System and the Israeli Company Elbit holds a key partner position in the UK Military Flying Training System Consortium. All of this from a standing start. Someone in MoD would do well to prepare for a backlash. Does political sensitivity or self-sufficiency ever feature in UK MoD procurement thinking and decision-making? No, of course not. Could the MoD not have found, encouraged or incentivised UK home-generated equivalent solutions? Yes of course it could have done that, and should have done, but it chose not to.

UK MoD Procurers are too lazy and too short-sighted to operate any meaningful form of National Defence acquisition strategy, so rivers of Taxpayer's Arms Cash flow out of Britain to overseas Suppliers. It's also shocking that many Israeli Defence Technologies and Systems are more advanced than their UK equivalents, considering the vast sums of UK Taxpayer cash the MoD has blitzed through in recent years, and continues to blitz through today. Israeli Defence Procurers with much smaller budgets, demonstrably get significantly more 'bang for their buck' than MoD's Bristol-based expensive and plodding 'Defence Equipment and Support Organisation' (DE&S) and the rapid penetration of the UK Defence Market by Israeli Companies proves the point. Students of the Arms Cash world have watched this happen, and it has stoked immense curiosity and speculation.

In the pre-Gaza incursion period, successive UK Governments started to align more and more closely with Israel, despite all the human rights and corruption issues which pre-dated those events. UK Governments effectively threw the genuine political aspirations of the Palestinian people, as distinct from Hamas, under the geo-political bus some time ago, so UK continues to export lethal UK Defence products to Israel today. It was shocking to learn that UK Forces were flying surveillance missions over Gaza during the Israeli invasion of the Strip, so the UK Government is not simply diplomatically tone-deaf it seems, but intentionally confrontational also as it was arguably a co-participant in the Gaza Strip invasion by Israel.

Abel and I had progressed as far as we could in Paris, and very surprisingly to me, I obtained Board-level clearance to proceed further, so now it was time to go to Tel Aviv. It would be my first visit to Israel, and I had mixed emotions about that. I'd imagined what it might be like of course and I landed at David Ben Gurion Airport fizzing with all the excitement which travelling for the first time to any country, or para-military State like Israel brings, especially where the pre-history of that country is so rich historically. I confess to a sense of apprehension too. Israel's Defence Industry reputation is well known internationally for several reasons, but perceptions and reality make strange bedfellows sometimes; I needed to experience it personally and make up my own mind, but I remained ambivalent about Israeli politics as many people are.

After queueing patiently in a long line of passengers I reached the Immigration Desk and handed over my Passport. For those who don't know about such things, some UK Businessmen are allowed to hold two British Passports if they need to travel through or to countries of conflicting political persuasions; India and Pakistan for instance, or Saudi Arabia and Israel. I did hold two Passports. I kept all the Islamic stamps in one and everything else in the other as I did indeed travel on Defence and Military Aerospace business both to Pakistan and India, and now to Israel and Saudi Arabia, as well as to many other Arab and Eastern European customer States with extreme political differences. Equally, I'd also been whisked out of UK once on an RAF plane to Saudi Arabia, and thence into Kuwait, without once using a Passport. Needs-must sometimes.

The Airport Official took my Passport and she looked through it very carefully. This wasn't the cursory flick through the pages I usually experienced in many airports, and without looking up, she asked if I wanted a 'State of Israel' Stamp in my Passport and I replied no, so she told me to stand aside from the queue and handed my Passport to another Official, who asked me why I didn't want the Stamp. I explained how I needed to travel throughout the region and the possible difficulties I might encounter with the Israeli Stamp in my Passport. She knew pretty much exactly what I was going to say before I said it, having doubtless heard it all a thousand times and more before.

A third Official appeared and ushered me into an Office for an Interview, where he asked me what the purpose of my Visit was, and I explained that it was a Business trip, Aerospace Business. He wanted to know who I was going to see and what we were going to discuss but I told him it was Defence Aerospace business and I couldn't discuss it with him. I gave him Abel's number. He left the Office to make the call, returning about fifteen minutes later. I was clear to enter Israel and he wished me a good stay. On my way out at the end of the trip they took me apart.

Abel met me as I exited the airport and I reflected to myself that it was good to have an established legitimate, overt local Business Partner for a change. It was a relief not to have to be concerned about local Agent approaches or Arms Cash issues. If there were to be any of those it would be Abel's problem not mine, and the Israelis already had enough of their own to last a lifetime.

A criminal investigation had probed the sale of German-built submarines and Corvettes to Israel resulting in seven arrests, including Thyssen Krupp's Agent in Israel, who later agreed to be a State witness, plus a couple of retired Admirals, including the former Deputy National Security Council Chief. The alleged crimes included bribery and money laundering relating to the deals, and the former Agent was sentenced to a year in jail with a $2.8m Fine. Here we have a fine example of both Arms Cash Agents and Influencers at work.

In another case, the sale of Israeli missiles manufactured by a Herzliya-based Company to Uganda was exposed in Der Spiegel as part of an international investigation into Arms Cash bribes. This Deal was characterised by a classic complex arrangement of Arms Cash Companies with overseas registrations, as with the Sri Lankan MIGs deal. Hensoldt, a German Company, supplied components to the Israeli Company which manufactured the Missile Systems, despite deep concerns within the Hensoldt Management team about the deal. One comment in the Der Spiegel Report reflected this concern by referring to '…the tension…' between the Sales team at Hensoldt and their Company internal Compliance Division. It was reported that Hensoldt had also supplied items to Saudi Arabia, despite an explicit Export ban.

In Zambia, leading Israeli Defence Company Elbit Systems, was involved in an Arms Cash deal, and four top Defence Ministry officials were arrested on suspicion of corruption in a $500m deal with the Company. Some years earlier it had been alleged that Bribes were paid in the Israeli Barak Missile sale to India, claims which resulted in a seven-year investigation, in which undercover operators claimed that high-level 'Influencer' bribes had been paid, as had Arms Cash payments to Agents. Elbit also experienced Arms Cash problems in Bulgaria. Thankfully, I wouldn't have to wade through pools of Arms Cash with Abel.

Before my first visit to Tel Aviv, I'd managed to convince myself that any business I might secure in Israel couldn't be classified as 'lethal' because I was marketing 'Defence Services' rather than lethal hardware. But I'd deluded myself on that point. So, here's a simple maxim for all Defence Company personnel and Politicians to absorb.

Whatever equipment, system, technology or services your Company makes, markets or sells, be it high-tech or low-tech, hardware, software or firmware, tangible or Intellectual in nature, they are all interrelated components in an overall national killing capability. As Suppliers you have no post-sale control over what you have supplied, regardless of what Contract clauses might stipulate, and the recent gifting of lethal products on a massive scale to Ukraine and Israel by the USA and UK highlights this issue.

Once sold or gifted and delivered to a Customer, any Defence Company's products have entered an uncontrollable zone and no amount of pre-sale bureaucracy or legislation can predict or control how, when and where Defence items will be used, nor against whom. Friends today perhaps, enemies tomorrow maybe. The outcome of your sale is an enhanced national ability to kill, and everyone who works in any Defence Company, from the Mail Room staff to the CEO, contributes to the process and shares the responsibility, as do the Shareholders, who sit back, hands-off, and enjoy the spoils of War; Shareholder Arms Cash. Let me say it once again just to be clear, once you sell the kit you have no control, and to believe otherwise is delusional.

My meetings with Abel, the IAF and other Companies during this first visit to Israel went well enough, and the Prospect started to coalesce. I flew to Israel several times to progress and convert these Project ideas into business. In Tel Aviv I usually stayed in one of the Dan Hotels, nice enough, but so bland, and I believed and acted as if no form of electronic communication was ever secure in-country. I attended meetings with Companies and with Israeli Air Force Officers, one of whom I'd met before in Romania.

In-between official appointments Abel was a superb host and a very well-practised Guide, but even during a visit to Caesarea to see the impressive archaeological excavations, he did not let the excitement of a mere Roman civilization stand in the way of an excellent seafood lunch, this time in a small Café which overlooked the Excavation site. Afterwards, Abel took me back to his home and showed me around the local Districts and perchance my visit coincided with Yom Hazikaron, Veteran's Day; an interesting juxtaposition for me, a former Officer in an Islamic Army.

Most Israelis observe Yom Hazikaron as the British used to observe Armistice Day, i.e. properly, and just about everyone in Israel participates. Abel and I had been driving along a major road just before the appointed hour when all traffic came to a halt. No one told the drivers to stop, they just did, and they got out of their cars and stood in silence, many with heads bowed. National flags fluttered from virtually every car; I kept one as a memento.

Back in Abel's village people gathered for an evening event where images of local fallen soldiers were projected onto a large screen, whilst relatives and friends shared their personal memories of them. Their testimonies gave real substance and emotion to the event. By way of contrast, I recall standing by my home War Memorial in a little Dorset village in watery sunlight on a cold November morning, and just as the final notes of *The Last Post* faded into the thin frosty air, an impatient Supermarket delivery truck driver, annoyed at being asked to stop, revved past the little Gathering at full throttle, drowning out the notes.

I've served in two Armies and lost friends in both, but I'm not sentimental about Remembrance ceremonies, as I remember my friends at any time and often do. Some people do opt to 'die for their country' but mostly, men and women don't sacrifice themselves, rather, they are sacrificed by the ineptitude of Politicians and Generals. Those hundreds of thousands of young Conscripts who were marched to their deaths on the Western Front were certainly sacrificed. Soldiers go where they are told to go and do what they are told to do when they get there, that much is the same the world over.

After more recent Conflicts, the difficulty has become balancing remembrance with the thirst for revenge; where is the cut-off point? Technology makes speedy revenge accessible and repeatable, and one by-product of the modern Defence Industry is the gift of grief to even more people. Bungling Politicians might light the fuses which inflame Conflicts, and inept Generals may send men and women to their deaths in the field of Combat, but the Defence Industry gleefully provided them with the enabling mechanisms. A crack of conscience opened-up in my de-sensitised corporate psyche.

Some of the discarded 'enabling mechanisms' of War, in the form of Tank hulks, adorn the road between Tel Aviv and Jerusalem. They rest there idle now, silent and impotent, standing as permanent monuments where they came to an enforced halt many years ago. Ironically, Tanks rest similarly in Russian towns and in Ukraine too, as monuments to 'The Great Patriotic War' (a phrase now banned in Ukraine). Once, these Israeli Tanks were camouflaged, but now they are highlighted and painted vividly and preserved for all to see and remember, the tangible memorials of conflict past, and if ever there was a city with a 'Past' it is Jerusalem. Abel and I pulled into in a car park overlooking the Mount of Olives and got out to survey the City, but despite being a first-time Visitor to Jerusalem, I already had a mental graphic reference to the ancient city from which to draw.

A large Victorian chromolithograph entitled 'The Destruction of Jerusalem by Titus' hangs on my Study wall at home, and the viewpoints of the Victorian Artist David Roberts then, and mine now, were similar. The great picture depicts Roman Armies streaming down across the valley and up to the walls of the flame engulfed City. Roberts' magnificent huge painting was sold in Rome in the Sixties and has not been seen since, and my original chromolithograph was created by Roberts' Belgian Plate Maker, Louis Haghe.

Now for the first time I saw this great walled City before me, and with Roberts' epic painting already imprinted on my memory, this view of Jerusalem felt familiar. Before we set-off into the heart of the City though, Abel was anxious to establish if I was religious. This was an important question for practical reasons rather than religious ones, since Abel conducted two types of walking tour for his business guests. Those of a religious disposition experienced Abel's whistle-stop tour, whilst the irreligious, a category which included me, experienced a much longer and more satirical version of the Holy City's history. Just to set the tone I told Abel my most irreligious joke and he reciprocated; that did the trick and off we went.

Abel pulled no punches with his personal interpretation of historic events as we walked through Jerusalem; the Gospel according to Abel one might say. He described to me the 'Jerusalem effect', that unearthly phenomenon experienced by the religiously inclined, or by those vulnerable to spontaneous spiritual conversions, who having visited Jerusalem were apt to gaze out of aircraft windows and see accompanying Angels. Such is the Jerusalem effect. Eager to experiment I tried this for myself, providing opportunities to all faiths represented before me to exert influence. I visited the Holy Sepulchre and the Garden of Gethsemane, as well as The Dome of the Rock and the Wailing Wall; I bought souvenir artefacts from each and every denomination; I smiled at Nuns, I nodded at Priests, I greeted Imams and Rabbis. But not a solitary

Angel of any persuasion did I see. I was and I remain surely beyond redemption, and I accept my fate.

I did make a point though of honouring my absent Arab friends by visiting the al Aqsa Mosque on their behalf as promised. On a subsequent Jerusalem visit, when accompanied by a badly lapsed former Catholic altar-boy colleague, an ex RAF Officer who was making his first visit to the Holy City, I did witness the Jerusalem effect at first hand. My archetypically tight-fisted Yorkshire friend put his hand in his pocket to buy Rosary beads for some aged Nun his wife knew, and when a Yorkshireman spends money like that you can be sure that something pretty supernatural is going on.

I'd already guessed where Abel stood on the sliding scale of religious fervour when we were driving to an appointment with Elbit Systems. We encountered (separately) two hitchhikers, the first of whom was a young soldier. Abel stopped immediately to give the young man a lift, making a diversion to get him where he needed to be. But when we later encountered a hitch-hiking Haredi Jew, Abel didn't stop, and was instead dismissive, because the Haredim oppose personal service in the Israeli Defence Forces. In Abel's own words 'they preach but they don't practise' and as a military veteran, that grated with Abel.

We wandered around Jerusalem trading religious jokes, but we weren't guilty of discrimination, since between us we would probably have upset all Faiths. As we walked, the constant 'in your face' presence of the Israeli Security Forces was an ongoing, intimidating physical reminder of the fragility of peace, and of the para-military flavour of daily life for all Jerusalem's citizens; c. 20% of Israel's population is Muslim, whereas some assert that 80% of Palestine's population is Jewish. Such is the nature of the challenge.

A Humous snack shared with some of Abel's Muslim friends in the hills outside Jerusalem, preceded our drive to Jaffa, and Abel had a practical culinary reason for going there. He drove into Port through the commercial dock entrance and parked, or rather, abandoned his car on the Quayside. Great timing. A small open fishing boat piloted by a friend of Abel's came alongside just as we walked onto the Quay, and after an exchange of greetings, Abel gestured to his friend and selected something from his catch. The fisherman gathered up the selection of fish in a basket, which he carried to the small café on the Quayside; the owner was yet another of Abel's many friends, so it was bear-hugs and greetings all round. Sitting on the little Quay beneath a sun canopy we enjoyed a selection of the freshest fish, served up by Abel's Cafe-owning friend, who joined us to drink wine and gossip as the Mediterranean sun set. It was one of those rare unhurried evenings; simple seafood, wine, and worldly people who were at ease with each other, and who were able to express and exchange differing views without vitriol or malice.

There was no such ease when I got back to the UK though, for despite being content for me to head off to Tel Aviv and engage with the Israeli Air Force, who were very welcoming I might add, Main Board Members had scared each other and got cold feet. They performed their standard about-turn trick of chickening out at the vital moment. I was losing count of how many times this had happened. Despite establishing a great position in a potential deal with an equally great local Partner Company, I had to drop Abel and my new Israeli friends and pull out.

The experience highlighted yet again just how insipid Defence Company Boards had become. Companies which had splashed the Arms Cash around so freely in another era, had become gridlocked by fear of the 'new' Corporate Compliance and Risk-aversion mantras, to the point where they were unable to execute virtually any new international business opportunities. The new breed of Plc Director had been selected for passivity and compliant personal attributes it seemed, as none of them would ever 'push back'. There was no vision, no worldliness, and no hunger for business. Senior Executives had become feather-bedded by huge bonuses for the soft MoD Contracts which their Companies would have won anyway, and for which they did not have to exert themselves one iota. The business strategy had become 'don't rock the boat'.

I finally realised that these folks simply didn't actually know how to do this sort of business, and that people like me really do scare the pants off them. If a business opportunity didn't arrive in the shape of a gift-wrapped UK MoD Tender, they couldn't handle it, even if it bit them on their fat arses. Well, actually, they mostly have slightly less-fat arses these days, because their wives ('Partners') make them eat Quinoa and take 'Spinning' classes.

Good, legitimate export business opportunities with not a cent of Arms Cash in sight are declined as a matter of course. What on earth would their Shareholders say if they understood that they're being short-changed on a macro scale? Probably not much if the Shareholder in question is a Foreign Bank or Financial Institution. An Israeli Company later picked up the blueprint for the project and won the business with it, using exactly the same methodology. If I had a Pound for every time I've heard some gormless UK Company Director say, on reading a Press Release announcing yet another win by an overseas Supplier, 'I wish we'd done that', then I'd be a very rich man.

Not only do today's Defence Company Directors have a very limited grasp of geo-political history, but they've also clearly been relieved of both testicles. Risk-aversion has become an Art-form. I've worked for several major Defence Companies in which the business strategy is articulated as 'don't lose anything we already have' rather than 'let's go find and win something new'. It hardly represents ambition, enterprise, or the will to win does it? I'd lost the chance to get back to Israel on business for a while, in fact I returned only on vacation years later to visit Mount Carmel and the stunning Crusader Castle at Acre, but in a demonstration of Business Developer's undiluted masochism, I surrendered next to the exploratory lure of the *'Jerusalem of the East'* and took flight to Kiev. I could never have imagined the 'Road to Damascus moment' which awaited me there.

Postcards from Kiev and Kharkov (pre-Invasion)
It seems ironic now, that in those distant days which filled the vacuum between the Orange Revolution and the Russian invasion, I accepted that Defence business outcomes in Ukraine would be harder to achieve and less rewarding in the short-term, than any achievable in the Middle East. How things change in a few short years. These contrasting Regions, though strangely connected in so many ways as I was to discover, remained worlds apart in others, whilst Kiev's cultural appeal was in an altogether different and higher league than all the Gulf States combined.

The business rationale for visiting 'broke' Ukraine at that time lay in recent history. The fall of the Berlin Wall in 1989 and the then receding Soviet (!) threat caused dramatic shrinkage in Western Defence procurement, and heaven forbid, many people even clamoured for a 'Peace Dividend'. But Defence Industry Executives consoled themselves with the prospect of compensating for this reduction with the lucrative deals they believed would now be available to them in Eastern Europe. The former Warsaw Pact States had been equipped with Soviet weapons and systems, which whilst robust in nature, lacked the technological sophistication of their Western counterparts, and this technology gap widened rapidly in the digital era.

The technical differential was limited in respect of basic equipment such as Small Arms, where Soviet-designed weapons were as robust and effective in combat as most western types, but at the higher, more sophisticated levels of Aircraft, Command and Fire Control Systems, Missiles etc., the gulf was wide and becoming wider. When former Eastern-bloc States declared their aspirations to join both NATO and the EU, Defence Companies swooped down on them like ravenous birds of prey.

Knowledge of Eastern European Defence procurement protocols and business practices were generally scant amongst Western Defence Companies, so reliable local business partners and contacts had to be identified and developed from scratch, which created scope for 'entrepreneurial' approaches both ways. In short, it was an Arms Cash Runner's jamboree.

Supplier excitement over this mountain of potential business was tempered though by concerns over accessibility and the ability of Eastern European States to pay for all this expensive stuff. In a very bad case of the 'pot-calling-the-kettle-black' corruption was frequently cited as a key concern by Western Companies, particularly regarding Ukraine, and several times I had been told by Executive Boards not to go there.

The Eastern European market was ripe for exploitation by Agents, Influencers and their Runners, all of whom got to work swiftly. I embarked on an Eastern European mission-bonanza to promote and market a whole variety of Western Defence Systems and Aerospace Services; basically, I cold-called Eastern Europe. I could hardly wait to get into the Region, as it made such a refreshing change from the over-worn path to the Middle East, and most of the newly targeted countries were of great historical and cultural interest and cried-out to be visited and explored. There are only so many times that a walk through a Dubai Shopping Mall excites anything other than terminal boredom.

It had been impossible for me to visit many of these Eastern European States during my MoD days, because the Ministry maintained a list of forbidden countries where Staff visits were forbidden; the 'Red List'. Permission had to be sought for some regional visits but was often declined, whilst other States were simply taboo. MoD was paranoid about security during the Cold War and with good reason, as the Spy-fests of the Fifties, Sixties and Seventies had demonstrated how useless UK Intelligence Services were at catching Spies or keeping secrets. Britain's key alliance with the USA came under significant and enduring stress at times during these years, so MoD's Security Officers became increasingly diligent; not necessarily more effective, just more diligent.

The visiting MoD Security Officer responsible for my discreet little Outstation MoD Office also served as a Methodist Lay-Preacher on his days off, and with his shock of snowy white hair he certainly looked the part. Equally, he could have passed for Santa Claus. He visited our nondescript little Office periodically as part of our ongoing Security training programme; perhaps 'Indoctrination' would have been a more appropriate term. He delivered presentations and briefings about the latest Eastern-bloc threats and tactics, and the part of his visits I enjoyed most were the made-for-MoD Security films. I particularly remember one called 'Eastern Approaches', not to be confused with Fitzroy Maclean's wonderful book of the same title, which highlighted how the Soviet Union and its Satellites approached and 'turned' UK Defence Staff.

I became well versed in these espionage case-histories and could recite a litany of the enemies of the State. I knew everything about Harry Houghton, the Portland Spy Ring, the Krugers, Lonsdale and the escaped Master-spy George Blake, plus the uber-treacherous Philby, Burgess and Maclean triad and that pompous idiot Blunt, and all the rest of them. In this new post-Gordievsky era, I now had the opportunity to visit and see some of these former Cold War adversary countries for myself, the very ones I'd been prohibited from visiting as a young MoD Executive all those years ago.

The Ukraine was especially interesting to me, owing to its comparatively recent separation from the USSR. Competing Ukrainian factions either celebrated this event, or conversely regretted and vowed to overturn it. Strange though it may sound now, not everyone in the newly 'free' Eastern Bloc wanted to align with the West or regarded their newly independent situations as positive, and I heard exactly these sentiments in several countries at that time.

Newly independent Ukraine sought to demonstrate its determination to stay free of Russian influence though, although this was always going to be a tough call, and intense political infighting continued. As Ukraine started to accept and encourage Western business visitors, I booked a seat on a Ukraine International Airlines 'plane and headed East.

Pre-departure, I'd cast about as ever, to identify some connection which I could make, however wispy, with my target country and with Kiev in particular. The best I could come up with was my childhood memory of Mussorgsky's 'Pictures at an Exhibition' and specifically of 'The Great Gate, the Golden Gate of Kiev'. Mussorgsky's music echoed nostalgically and evocatively in my head; was there still a Great Gate to be seen I wondered? I hadn't researched it, having only just returned home from a Singapore Special Forces Project visit, but I was eager to find out.

My Passports had been taking a beating and a further Passport challenge awaited me on arrival in the 'Jerusalem of the East' as Kiev is known; David ben Gurion Airport déjà vu swept over me; there must surely be some secret international association of Immigration Officials, akin perhaps to the Concierge network of the Grand Budapest Hotel. The Immigration Official in this instance was a tall, strong-looking blonde lady, and perhaps at the time, Western Executives still had a novelty value.

She asked a whole battery of questions and wanted to know the reasons for my visit. She was not content with any of my answers, so clutching my Passport in her vice-like grip, she marched me off to an Interview room and closed the door. Now it really was a case of déjà vu. There were just the two of us in the room and a whole variety of scenarios ran through my mind.

She went through my small carry-on bag with a fine-toothed comb, shaking out the contents vigorously onto a table and picking up every single item in turn to examine it minutely. I didn't know what she was looking for, so there wasn't much I could say. Eventually, she tired of the game and off I went into the city to find my Hotel. The shadow of Arms Cash was never far away in Ukraine though, as demonstrated by the case of the Sri Lankan purchase of Ukrainian MiGs.

In another strange episode, a Ukrainian registered aircraft had made an emergency landing in Kano in Nigeria and was discovered to be carrying around eighteen crates of ammunition and Mines; the plane was impounded initially but departed abruptly days later. Mines are especially controversial munitions and not simply because of all the past Princess Diana hubbub. I'd once spent four months operationally deployed in a mined area where I'd developed a healthy respect for Landmines, and later, whilst working for another major Defence Company. I'd championed the development a Humanitarian de-mining system, and briefed the EU's Humanitarian Commissioner, Emma Bonino, about the project's potential, sadly to no ultimate effect, as that project was lost in a Defence Company takeover, which says a lot about Defence Company priorities.

The issue with the unscheduled Ukraine-Nigeria Land Mines consignment, was that apart from being used in their design-form either as anti-tank or anti-personnel mines, the larger anti-tank mines are also constitute conveniently pre-packaged bulk explosives. Mines are modular and thus are easy to move around and to store or hide, but they can also be disassembled, divided-up or bundled into lethal packets, or simply modified to provide the deadly key ingredients of Improvised Explosive Devices (IEDs) which were so lethal in Iraq and Afghanistan. Anti-tank mines configured singly or in multiples as IED's, and sometimes used with anti-personnel mines as low-pressure detonating devices, can produce massive explosions capable of destroying the best-protected armoured vehicles. Thus, even discounting the humanitarian issues created by using Mines in civilian areas, any unscheduled planeload of Mines, such as in Kano, will attract close and urgent attention.

One Media report claimed that the aircraft had been carrying munitions purchased under a contract placed by the Defence Ministry of Equatorial Guinea with a Cyprus registered Company called 'Infora'. H'm. The Ukraine Government denied that the cargo belonged to Ukraine itself, although the aircraft was undeniably registered in that country, and there are not many civilian owners of bulk stocks of Land Mines. There was great sensitivity in Nigerian government circles owing to speculation that the cargo might have been intended for militant groups in the Niger Delta.

Widespread concern that Arms from Ukraine were ending-up in undesirable locations, reportedly focussed the British Foreign Secretary of the day on British 'Broker' involvement in deals emanating from Ukraine. Ukrainian agencies UKRINMASH and UKRSPETSEXPORT had previously been involved in the sale of weapons to Serbia, and potentially combining this rich source of Arms with British Arms Cash Broker know-how would be a marriage made in heaven. A visiting delegation of British MPs was presented by the Ukrainian deputy foreign minister (then Oleksandr Gorin) with a list of alleged UK Arms Brokers. Ukraine certainly enjoyed an international reputation as a source for illicit Small Arms to international conflict hot-spots and Ukrainian Government internal investigations identified multi-millions of dollars-worth of missing military materiel during the 90's. Someone might want to take another look at that, now that Ukraine is knee-deep in gifted Western weapons and munitions.

Whilst regular Western Defence Clients, especially the Gulf States, enjoyed a surfeit of cash which allowed them to indulge their appetites for increasingly high-technology US Fighter jets and other state-of-the-art Defence Systems, the opposite applied in Ukraine because the country was palpably broke. How ironic that seems now. One Ukrainian Air Force Officer confided to me that a shortage of cash for fuel kept Ukrainian pilots out of their cockpits and their aircraft out of the skies. The Western, well US geo-political strategy really, had been to sign-up as many ex-Warsaw Pact States as possible and advance NATO 'Asset deployment' as far onto eastern soil as possible, in the form of Bases, Missile platforms and Training missions, and prior to the invasion, NATO troops had trained in Ukraine. Did anyone stop to think about the impact of all that on Russia, No, of course not. Was it intentional provocation? Personally, I think so.

Western policies remained demonstrably confrontational and rooted in the Cold War mindset, despite the erosion of the previous Soviet Union threat, which momentarily gave flickering hope that something different could be achieved. Sadly, there were no Western political initiatives to optimise the situation, just a continuation and extension of NATO expansion with all the negative effects that brought. I agree with the point of view which holds that these Western actions and gestures simply taunted the Russians unnecessarily and raised the diplomatic temperature for no good reason. It was an utter failure of Western political imagination, creativity and engagement. Years after my first visit to Ukraine, the events in Crimea and the later invasion of eastern Ukraine highlighted Russia's response to this provocative Western policy, and in my own opinion there was no need for it. British Foreign Policy diminished qualitatively to the point of invisibility in the hands of politicians with the attention spans of gnats, and an addiction to soundbites in lieu of any actual geo-political understanding. Palmerston must shudder regularly in his grave.

One Western Defence Company goal was to addict Eastern bloc States to the Western Defence technology 'must-haves' which alone would sustain their NATO interoperability, and pay for their NATO Membership cards. For Western, mainly US Defence Companies, it was and still is, the gift that keeps on giving, since once a State procures a high-tech Weapon or Aircraft System it must keep-on paying for expensive upgrades, technological refreshes and enhancements. It's a bit like Microsoft Windows for weapons in that respect; there's always a better version coming down the line.

My own business mission was a bit different and quite simple really, I wanted to find indigenous Companies with which we could team-up to offer international services, such as military flying training etc., and I was operating in an open contact mode, not diving around in the shadows like the hotch-potch of Arms Cash brokers who visited Ukraine, looking for Arms deals.

It was so refreshing to diversify from the Middle Eastern Arms fleshpots to which every Carpetbagger and Arms Cash chancer has beaten a path, although the Customers there had changed significantly since I'd started visiting that region, and it was less enjoyable than it used to be. I didn't expect to stumble over crocks of gold in Eastern Europe and certainly not in Ukraine, but there were compensations, and this initial early visit was just exploratory. Maybe I could conjure-up some co-operative deal within the Aerospace sector in which the Ukrainians had a great track record; the 'Heavy-lift' Antonov's for example, made a big impression in the West. Ukrainians love Aviation, it's in their blood.

I started my visit programme with a meeting in Kiev with a Joint Stock Aerospace Company, which had a mix of private and Governmental shareholders. Executives there were pleased to meet-up and discuss possibilities, but although they were eager to get started on something, we (i.e. me) had to come up with all the original ideas. My Company's debilitating and laborious commercial processes, dense and mountainous as they were, were ever a killer though, so we were too slow and too plodding to respond even when good opportunities were identified. Our Commercial teams were entirely inward looking and regarded export Customers as an complete embuggerance.

During one aerospace project meeting, a Ukrainian Company Operations Director asked me 'what happens next?' and when I'd explained our Company internal 'processes', our potential Partners didn't understand or recognise any of the steps I described. Instead they looked sideways at each other, confused and bemused by it all, and asked me why it took so long to get decisions on such simple matters. Frankly speaking, I shared their incredulity and I still do. No, that's not strictly true, I gave up. It was pointless.

This first meeting in Kyiv started at nine thirty and ended two hours later, at which point Vodka was brought out. A series of toasts were proposed and drunk, then another bottle appeared and was also opened. It was apparently unlucky to leave a partly full bottle so we finished-off this second one too. Thus, was set the pattern and tone for the whole visit, a bit of chat, some good ideas, congratulations all round, then Vodka and more Vodka. I really couldn't handle too many meetings in a single day in Ukraine, my liver couldn't cope with it, but other meetings followed in rapid succession during the visit. I think I should be compensated for sacrificing my Liver on Defence Company business. I certainly sacrificed my sanity.

I took time-off to look around and I city-walked Kiev, that city involuntarily and indelibly imprinted upon me as a boy by Mussorgsky, for which reason I headed-off immediately to see the reconstructed Great (Golden) Gate of Kiev. There really is one. The reconstruction was impressive, and I was intrigued to discover that before his musical career blossomed, the young Mussorgsky served as a Cadet in the Preobrazhensky Regiment of the Imperial Guard, which boasted a distinctively shaped cap badge. That badge emblem was of a very similar design to one belonging to one of the three Infantry Regiments in which I had served. It was a modest connection 'twixt me and Mussorgsky, but it would do for now. I adore his music, so a wispy connection with him was good enough. Kiev, Mussorgsky and I were obviously soul mates and I progressed on that basis.

I crossed the Dnipro several times to see both sides of the city which straddles that great river, and I wandered around Kiev's great curving Besarabka indoor market, a veritable Albert Hall of a market, or so it seemed, from which the people of Kiev purchased meat, fish, vegetables, flowers and just about everything else. The most wonderful fresh flowers were available even in the bitterest cold. Ukrainians love flowers, they are as important in the Ukraine as is Fashion, and I was reliably assured that given a choice of spending their remaining Hrivna either on food or fashion boots, the young women of Kiev will buy the boots, but I've yet to put this to the test.

Impressed though I was by the reconstructed Great Gate, I was doubly impressed by 'Rodina Mat', that towering (over three hundred feet) Soviet (though not called 'Soviet' anymore) stainless-steel statue of 'Mother Motherland' now renamed 'Mother Ukraine', which dominates the Kiev panorama. Rodina holds aloft her trusty though sadly truncated sword, which is still c.16 metres in length/height. The sculptors apparently ran out of raw material so cut short the design.

Rodina's shield originally bore the emblem of the Soviet Union which was replaced by a Ukrainian emblem in August 2023. Ironically, Rodina faces East to welcome the liberating Red Army, the local history of which was reflected in the 'Museum of the Great Patriotic War' which Rodina Mat, now 'Mother Ukraine', marks for the Visitor. I wanted to go there but wasn't ready for the life-changing impact that was to follow. (Note: In Ukraine the term 'Great Patriotic War' has also been outlawed and the Museum's name has been changed too).

The human cost of 'The Great Patriotic War' remains almost beyond practical calculation and belief, as in modern times it's difficult to grasp the magnitude of that titanic struggle, unmatched in human history as it is. Those desperate feral battles with German forces for the great Ukrainian cities had been epic in scale, loss and strategic significance, and the Museum above which Rodina Mat/Mother Ukraine stands sentinel, provides a breathtaking and emotional insight into the most brutal and visceral of Conflicts. Defence Company personnel should be compelled to visit this place if they want to understand the consequences of their work, brought chillingly up to date now by current events. Modern conflicts such as Afghanistan, though deadly of course, pale into military insignificance compared to the ferocity, sheer human scale and desperation which characterised War on the Eastern Front.

The base of Rodina's column is a dome-shaped Memorial Hall, and on entering this base and looking upwards you can see the names of the patriots who fell defending their City. There are thousands upon thousands upon thousands of them in the ceiling. The galleried museum complex lies discreetly and passively within the hillside adjacent to the dome, ready to stun the visitor with chilling exhibits, not least of which was the huge transparent Perspex cube into which have been tossed, contemptuously, Iron Crosses ripped from German Soldiers. These military Decorations once so sought after, once so envied and worn with surging pride, lay deposited and discarded now, consigned to this plastic waste-bin of history for all to see and disdain. But a truly heart-stopping exhibit ambushed me as I weaved my way through the galleries. I wasn't sure what it was at first, since part of it was obscured and the light in the gallery was subdued, but as I got closer it became horribly obvious that it was a German Army Field Guillotine.

This timber-framed apparatus held the characteristically angled steel blade which is raised by a rope and then dropped to decapitate a victim. Victims were strapped to a wooden board for the convenience of their Nazi Executioners who slid them along the platform and into position without the distraction of a struggle. This guillotine was a sophisticated well-engineered device, not an improvised local item. To the side of the gallery a large monochrome photograph bore testament to the operational use of this hideous device against Partisans; it was detestable both in concept and in operation.

I gazed at it for a while, transfixed by the horror of it, but gradually it occurred to me that if, for whatever reason, you feel the need to execute someone in cold blood in the course of a War, even during a campaign as unremittingly bitter and quarterless as the Eastern Front, then the quickest, easiest and most efficient way to do it, and in economic terms the cheapest too, is to use a simple bullet. Serbian Forces took great care to record on film their own merciless mass application of this technique in more recent times.

But if instead, if you choose to take the time and trouble to design, manufacture and to use an elaborate device such as this mobile Field Guillotine, then clearly something else is on your mind. The device was a calculated articulation of terror and sadism. It was a deliberate and conscious effort to inflict the maximum degree of horror and humiliation upon both the victims and the innocent witnesses to its use. The operators of these devices must have pleasured in their use and effects and I stared at it for a few moments, and then, I really don't know why, I felt the urge to touch the wooden trestle and I experienced a disturbing chill as I did so.

It was only then that I noticed the 'Label'. This was a metal manufacturing plate or label secured with screws and stamped with manufacturing data. The device even had a 'Mark' Number, this particular version was a Mark 3, I think. Whatever happened to Marks 1 and 2 I wondered? How and why did someone realise that a modification had to be made to improve performance? Did 'User-feedback' prompt a design enhancement? There must have been a manufacturing Data-Pack which enabled the serial production of these devices and a logistics plan too for their distribution. I was rooted to the spot by its ghastliness. Then it clicked. I made the connection between the Field Guillotine Mark 3 and me.

This Guillotine was a product of the Defence Industry of its day; someone conceived and designed it, somebody ordered it, somebody manufactured it, somebody sold it, somebody paid for it, somebody used it and finally somebody died on it. It caused horror and fear and resulted in death. Years later, here I was looking at it during a gap in my own defence marketing schedule of meetings, during which I promoted some of today's Defence Products, Systems and Services. Someone like me maybe had sold the German Army this Guillotine. Was there any difference between us? I bet we could both justify our actions to ourselves. It was a transitional moment which hit me like a train for many reasons.

There had always been 'two of me' travelling around the world on Defence Business. One was the desensitised, mission-focussed ex-military Defence Industry Executive, and the other? Well, someone quite opposite, someone I kept locked in a box; I kept the two apart. Like AC and DC current they weren't compatible. But in the Kiev Museum of the Great Patriotic War, or whatever Ukrainians choose to call it these days, these two characters confronted each other across the Field Guillotine, and I froze emotionally.

I never envied all those dull Commercial and Technical folks back home who were safely insulated from this sort of stuff and I considered myself to be the fortunate one. Defence Company activities and the so-called 'Corporate Offerings' are sanitised, and the whole language, the argot of the modern Defence Industry has changed, modified and softened to deliberately obscure the consequences of the use of modern Defence equipment. Terms such as 'collateral damage' the 'use of assets', 'surgical interventions' and so on, all constitute delusional jargon which insulate both Companies and Military forces from the horror of the effects they enable and unleash.

We watch our TV screens unmoved as high-tech precision-guided weapons are fired with unerring accuracy against pre-selected buildings thousands of kilometres away, even passing through nominated windows or destroying selected vehicles as they drive down some remote desert highway or street. We were 'wowed' by the technology at first, less so now, forgetting or failing to consider for even a moment that we have just witnessed someone's death, someone innocent maybe. Probably.

Computer games and Military digital reality have blended to the point where the differences and consequences are indistinguishable unless you are the victim. Perhaps I could show young Defence Systems Designers a picture of the Field Guillotine Mk 3 and tell them that the outcome of his or her own work could have the same effect on people; i.e. horror and death. What reaction would I get? Would they see the parallels but continue to march forward with the same digital gusto and enthusiasm? Would they pause even for a second to evaluate the consequences of what they design and produce, or are they now institutionally and technologically inured from such outcomes? Military Drone Pilots certainly are. Probably they would be curious but indifferent, a case of 'what's-that-got-to-do-with-me-syndrome?' Consider for a moment this extract from a speech by Dwight D Eisenhower:

Every gun that is made, every warship launched, every rocket fired signifies in the final sense, a theft from those who hunger and are not fed, those who are cold and are not clothed. This world in arms is not spending money alone. It is spending the sweat of its laborer's, the genius of its scientists and the hopes of its children. This is not a way of life at all in any true sense. Under the clouds of war, it is humanity hanging on a cross of iron.'

Elsewhere in Kiev I found alternative crosses and made less chilling but nevertheless spooky discoveries. I 'found' Pechersk Lavra, the Monastery of the Caves, where I descended into the labyrinthine bowels of the Earth holding a candle 'twixt my fingers to illuminate the path through the largest monastery tunnel complex in 'Old Russia' (yes, that's what it was called). The wispy candle flame threw an eerie pale-yellow light across the mummified corpses of the Saints who rest therein. In times-past Christians withdrew into these caves and tunnels to protect and preserve their faith from persecution; this was a place where religious hermits took refuge and where Saints remain preserved, their mummified hands protruding eerily from the shrouds which enfold their remains. No doubt we now have a missile somewhere which could race through this twisting turning cave complex, having discovered at Tora Bora in Afghanistan that we had none with such characteristics when we needed one.

I discovered too the beauty of St. Michael's golden domed monastery and St. Sophia's Cathedral, and who would not be struck by the visual joy of the city's architecture in the watery autumn sun, bathing as it did the golden reflective domes of the city. I fell in love with Kiev, the 'Jerusalem of the East', where Christians retreated into caves to preserve their faith.

Was it time for me to retreat somewhere? Not to preserve a belief but perhaps to escape one? Things had changed irreparably for me now and I knew it. There had been 'two of me', yes, the corporate deal-fixer, armour-plated by military service and focused solely on 'outcomes', untroubled by ethical rules; competitive, combative and abrasive. And the other guy? Well, he was still that little boy sitting on a riverbank in a Bluebell wood in rural Berkshire, untouched by the wider world and happy simply to hear birdsong and to watch freshly unfurled translucent lime-green leaves sway in Spring breezes.

Of course, that young boy was blissfully unaware of the Atomic Weapons Establishment which lay behind the high barbed wire fences atop the hill at the far end of the village in which he lived. The Mk 3 Field Guillotine had severed the rope which held him back and he came for me now. I was going to have to account to him for my journey from being him to being me. How on earth was I going to do that?

It was refreshing, revitalising even, to be in a proper city again after so much time in the Middle East, and Kiev was awash with cultural artefacts, exciting architecture and Art, lots and lots of Art of all sorts. Though rinsed through by Soviet ideology in modern history, Kiev remained a place which had been a crucible for great events, it was a City of depth and substance which demanded attention, exploration and concentration and which provided a stark contrast to the modern imposter Desert cities of the Gulf which I visited so frequently. Those

147

were Cities born without souls, the showcases of modern architectural indulgence and insanity, the articulation of wealthy exhibitionism masquerading as progress. No one walks very far in Gulf cities, they take short hops to and from Taxis, Hotels, Shopping Malls and Offices; pavements are few and far between and too hot to use anyway.

In Kiev it was good to city-walk again, even though I narrowly avoided being mugged soon after arriving, which was not something I'd ever experience in the Gulf, but which proved there are always two sides to any ledger. After several days of meetings in Kiev my Connections urged me on again and so I flew to Kharkov, which had a much stronger Soviet flavour about it, maybe hence the awful hotel I suppose. The resultant Civil aviation experience provided another metric for comparison with the rest of the world.

The internal flight from Kiev to Kharkov provided the next wake-up call of the trip after Vodka poisoning, the attempted mugging and the Field Guillotine in the Museum. Passengers gathered in a tight Arctic Penguin-like co-protective huddle in the darkness outside the drab Soviet-designed Terminal building, and we rotated in and out of the centre of the huddle in the sub-zero wind. When called forward eventually by a distorted Tannoy announcement, a spontaneous passenger stampede broke out across the airfield's concrete apron towards an ageing twin-engined Antonov. The rush caught me by surprise, just as I'd once surprised the RAF in Scotland by 'rushing' one of their operational bases when I landed unannounced with my troops in a Special Forces C130 to test their Base security, as part of a 'TACEVAL' (i.e. Tactical Evaluation) process. We were in and out in thirteen minutes on that occasion, but I recognised several common factors between that Scottish military experience years previously and Kiev now, insofar as there was a lot of concrete to cover at pace, it was freezing cold, and the aircraft looked a bit battered.

The reason for this Ukrainian passenger dash was not 'TACEVAL'-related though and it had a more practical purpose. Passengers with baggage, which meant almost everyone, headed at best speed towards the front of the aircraft where the Crew had opened a Cargo Hold door. Passengers had to pitch their own baggage up and into this Hold as no handling crew was available. It was a disadvantage to be physically short because the compartment was high above ground level. Baggage Basketball.

148

The winners of the trans-concrete race gained first access to the boarding steps and tussles broke out between competing passengers. On-board seating was a random process, insofar as those who made it up the steps first had a choice of seats. There was a good reason for haste in securing a seat as I soon discovered. I managed to grab a seat as did my travel companion, but when he reached for his seat belt, he found it to be devoid of fastening buckles, which had long ago become detached from the frayed webbing straps. At my behest he tied the free-ends in a knot which he'd learned in the Boy Scouts; I'm sure any BA Flight Attendant would have done the same.

The flight deck was separated from the passenger compartment by a flimsy curtain, red of course, which looked a little like a holed tablecloth tied on with string. Our Flight Attendant completed her pre-flight routine by dousing the cabin lights via a series of very long Bakelite rocker switches on a bulkhead, which would have looked at home aboard the 'Memphis Belle'. My favourite moment though was refueling. This took place from an old, slow-pumping fuel-bowser which pulled up beneath the wing as we boarded. I noticed that the earthing safety strip was too short to reach the ground and it waved unrestrained and fluttered in the chill breeze like a pennant. Aviation fuel splashed freely everywhere as we waited for the old pump to complete its task and to ready the aircraft for take-off. I sat back to enjoy the night flight.

Once airborne (ish) the seat-race winners enjoyed their relatively quiet seats at the front of the passenger cabin, whilst the less knowledgeable losers like me sat in over-wing positions and were deafened by the roar of the propeller engines. We were vibrated vigorously and temporarily deafened on take-off. As there were more passengers than seats, the surplus passengers endured the flight squashed into the cabin cargo area or sat cross-legged in the aisle behind the cockpit cloth. The bumpy and discouragingly low-altitude night-flight seemed to last an eternity until at last we touched down in Kharkov. I estimate that the whole experience took about ten years off my life. There were alternatives of course, we could have made the journey either by road or by rail, the latter option apparently took around ten hours and was not recommended even by hardened locals. Was that because *Red Grant* was hiding in a Sleeper compartment ready to pounce on Western visitors and strangle them, I wondered? Ah, the glamour of it all!

I'd hoped for a leisurely breakfast the following morning, but the hotel food was truly inedible, it really was, so even without a tight schedule there was no incentive to dwell over it. We were collected, tired and hungry later that morning by a car sent by our local business Host and the presence of the two black-combat-kit-clad Minders in the car set the tone for the rest of my time in Kharkov. The big Mercedes, also black, everything seemed to be black, had bulletproof glass, and the driver ignored most traffic lights without attracting any Police attention. We were driven around the city at Mach One speed. Surely this car could comfortably outrun the juddering Antonov which had brought me here.

Our Sponsor's offices were staffed with more of the black-clad brigade, and we were escorted everywhere by similarly attired Minders. Whether these Escorts were for our own safety or for the protection of the population wasn't clear. They drove us to the huge outdoor market which was set on a tract of wasteland, and it was striking not simply because of the sheer scale of the place, acre upon acre of it, but because of its diversity. The Market was arranged in themed areas, foodstuffs, clothing etc., but had surprising ethnic zones too. I hadn't expected to find a Vietnamese market in freezing Kharkov, and predictably, one could of course purchase anything that fits on, in, or around a Lada, and I daresay, an Antonov too if you looked hard enough. Such was the scope of the place. Market administration was run from a central brick building staffed by at least a platoon of the black-clad Minders who kept a tight grip on the whole operation. Kharkov seemed to be alive and buzzing in many respects, even if the City plumbing had yet to enter the current century. The human scenery was something to behold.

I didn't know if the young ladies of Kharkov, and I do not refer here to Hookers, were the product of some bizarre Soviet experiment or not, but they were of seemingly uniform height (tall), blonde, black-clad (what else?) and consistently slender and pretty. My companion observed that there must have been a 'mould' and I agreed as I watched a steady stream of them passing through the Park. Maybe the old Soviet regime had imposed a secret formula. My Translator, a local Newspaper Editor, told me that fashion was crucial to these young Kharkov ladies, they really would choose clothes over food and so it seemed, for demonstrably none of them was over-eating.

150

The next Kharkov phenomenon was musical. I was sitting in a Hotel garden taking early-evening coffee with my Editor friend (sitting outside still 'works' somehow despite the cold) when a young man appeared on the adjacent outdoor stage. Nobody took any notice and he sat and began to excite the keys of the outdoor piano. He opened with Jazz, but everyone ignored him, so after ten minutes of improvisation he changed tack and delivered Chopin. Still no one took any notice. When he eventually finished his 'set' I alone it seemed, applauded as he left the stage.

"No need to applaud..." said my Translator friend sardonically "...there'll be another one along in a minute or two..." and she was right.

For the next hour or so a stream of young musicians took the stage and played to a disinterested audience, which was fully pre-occupied with social chatter, coffee and vodka, whilst I was bewitched by the playing. I would happily have paid to hear music like that. Musicians, it seemed, like Mathematicians and Physicists too, were in plentiful over-supply from Ukraine's Institutes. At the time of my first visit years ago, my College Lecturer tri-lingual friend with a Doctorate was worth barely $350-$400 per month to the State. Some people in Kharkov were clearly making a lot of money though, but they were not on anyone's official payroll.

At the end of the week my Sponsor invited me to join him and his team of black-clad Minders at their regular end-of-month barbeque, the location of which was about an hour's drive away. The barbeque site was set in a small clearing deep in a huge forest which surrounded a lake. We drove down a long bumpy track through the trees to reach the place and the combined scene, lake and trees, was slightly Cumbrian in character, though on a massively grander scale. It seemed much quieter too with no habitation for miles around, and the uber-placid lake was disturbed only by incoming and departing ducks. The Team fired-up the barbeque, a halved oil-drum on angle-iron legs, none of those prissy 'Garden-Centre' models here, and 'the Boys' started to prepare fish whilst downing prodigious quantities of 'Beer Ukrainski', and of course, Vodka.

My Host offered me the use of a small rowing boat, which was moored in the reeds, whilst his team wriggled coals and gutted fish. I clambered into the small craft and pushed out into open water. When I paused mid-lake to look back, the woods were just as I'd imagined huge eastern European forests to be, very tall strong broad pines, scented, black and dark green and densely packed. The placid lake was disturbed only by my clumsy splashing oars as they sent out ripples, which were absorbed and smoothed to nothingness by the mass of water before they could even lap at the reed beds.

There was something extra special about this place and I understood why my new friends came here. What made it so special? It was silence. Without the splash of the oars it was completely silent. I looked up to search the sky and saw not a contrail in sight, not a plane to be seen or heard, no passing traffic rumble or human sounds. Far out in the lake I could hear neither the chatter of birds nor the gossip of the barbeque crew and I closed my eyes and bathed in that silence. My transient moment of tranquility was disturbed only by my Host, who bellowed across the lake and urged me ashore to share the cooked fish. My moment of silent isolation had gone, and I've longed in vain for another ever since.

A major focus of my first mission to Ukraine had been to explore Military Aerospace opportunities so I looked forward with great curiosity to my first meeting with the Kharkov State Aircraft Manufacturing Company. The Company had designed and manufactured innumerable aircraft types since their foundation in the 1920's and fascinating sepia photographs of Soviet Aircraft types adorned the internal walls of the Factory. It was truly a Plane-Spotter's paradise and I would be the envy of my aero-sexual work colleagues when I returned to UK. I would tease them with tales of the unique aircraft types I'd seen in the Factory; they were willing victims of course, masochistically longing to be tormented with Aero trivia.

Company Staff included many ex Royal Air Force personnel and as our offices overlooked a regional airport, these happy souls were in their element. Binoculars lay ready for action at all times on the window cills overlooking the panoramic runway-facing window, so every time an aircraft of even minimally passing-interest appeared, which meant virtually every aircraft from what I could tell, ex RAF fingers left their keyboards and wrapped themselves excitedly around the ever-available optical aids. As each aircraft made its final approach the dialogue was predictable.

"…ooh! it's a 237-4…" the first observer might say excitedly. Then a pause as others scanned the airframe.

"...no, no, it's a 237-5..." might come the challenging more knowledgeable and correcting response.

"How can you tell?" The riposte.

"...check out the rivet pattern on the port side of the forward cargo door..." another pause

"...ah, yes, yes...you're right; a 237-5"

The aeronautical trivia expertise of our resident 'ex-Crabs' was unsurpassable and their technical descriptions and mastery of aircraft variants were of legendary and suicide-inducing dullness. When the Red Arrows flew-in once I thought I was going to witness spontaneous group ejaculation.

There were no binoculars to hand on the Factory floor of the Kharkov State Aircraft Manufacturing Company though, and Workers noticeably kept their heads down as our Factory tour progressed, hosted personally by the Factory Director. Workers had neither the time nor the inclination to get excited either by our visit, or by the historic photographic record of their own Factory's historic output. Perhaps they lacked the critical 'rivet knowledge' which enabled ex Air Force Staff at home to rejoice in the aero-wonders of the world, or perhaps they were just there for the minimal wages they earned to feed their families in the tough post-Soviet economic climate.

Not everyone I met was glad to have seen the back of 'the Reds' and I'd heard similar comments in Leipzig shortly after the fall of the Wall. Some yearned for a return to the more certain and 'stable' (not my word) Soviet days when rent and food prices at least were constant. The rising cost of living was the price of freedom, and it was a timely reminder that seismic political changes in the East generally were going to take time to be absorbed and would not automatically be welcomed from all perspectives, other than those of the West.

The Factory Director was a serious aviation enthusiast and there was much to be proud of there. As he guided me through the AN-72/4 production line, I paused to look at a photograph of a huge 1930's 'Flying-Wing' type aircraft with umpteen propellers which looked like a fantasy aircraft from an old Comic or from a 1920's black and white movie. The Director joined me at the wall and peered intently at the picture with me; he tapped the picture with an outstretched finger and looked me in the eye at point-blank range. In a forgivably thick accent, he said in English

"Progress in aviation is paid for in blood…" I nodded but wondered whose blood he had in mind today.

After the Factory tour and various mini meetings, we adjourned to a Factory Canteen for snacks and of course Vodka Toasts. It was eleven thirty and my sidekick and I had other meetings to attend, but alas and predictably, the Vodka flowed freely. Toasts were drunk until my colleague looked ready to pass out. His brain must have been neutralised by the first five or six slugs of down-in-one-vodka which surely doubled-up locally as Antonov fuel. Remember the rule, once the bottle has been opened it has to be finished. With the Vodka gone, we all agreed it was time to go.

As a parting gesture the Factory Director arranged for us to visit the Company's outdoor Aircraft Museum, so a fume-belching Minibus with curiously patterned gingham interior curtains (many Eastern European minibuses seemed to have these little curtains, mostly in this same incongruous gingham patterned fabric, or similar) arrived to collect us from the grand front door of the factory. My companion slumped in a delirious mound on the rear bench-seat of the Minibus as we set off, chugging along and pumping out great blue exhaust clouds and acrid fumes. The Minibus moved slowly along the edge of Factory's vast runway, surely creating enough toxic emissions to neutralise the whole of the EU's renewable energy programme. My colleague floated in and out of his comforting sleep, leaving me to absorb both the full impact of the outdoor display and the fumes.

The Museum provided a comprehensive display of Soviet aircraft types through the ages, lots and lots of them. They were parked on a grass apron along the seemingly infinite concrete runway. I've never seen so much concrete. Our Hostess in the Minibus was surely related to Rosa Klebb; Klebb's mother perhaps, or maybe her stern elder sister. She commentated in consummate detail on each of the many aircraft types and on and on she went recounting national aeronautical triumphs, until two interesting things happened which woke me up a bit. The first was a morsel of Soviet Space history. Mamma Klebb said:

"…and this is the aircraft in which Comrade (yes, she did say 'Comrade' in the Soviet vernacular) Yuri Gagarin trained for his heroic Space Mission…"

154

Now she had my full attention. Those early Space Pioneers took huge risks and this aircraft had been flown to create the negative G-forces essential for Space weightlessness training. We're all familiar with this technique now but then it was innovative training for epoch-defining space travel. Sheldon Cooper would have loved it. It was worrying though as I was beginning to exhibit tell-tale signs of 'aero-sexuality' and my ex-RAF colleagues at home would be delighted.

But just as Rosa's Mum was finishing her Gagarin narrative, I spotted something even more interesting at the end of a line of parked 'planes. The 'thing' which caught my eye triggered a military memory. Surely this was something I'd first seen during my Rhine Army Service in an Intelligence briefing in Rheindalen, when I'd been shown a manufacturer's video of what was then a classified US weapon system. What was it doing here? The object of my curiosity was a missile-on-a-stick mounted on a plinth and I was fascinated by it.

"…that looks just like a Tomahawk Missile…'" I murmured, straining to peer between the little gingham curtains; I was destined not to get a closer look though.

"…and that concludes the tour" said Ms Klebb, abruptly pulling shut the gingham curtains and leaving me staring at the pattern instead of the missile on a stick. The chugging minibus wheeled around and took us back to the exit and not a moment too soon, for my companion was now in state of semi-consciousness, surely enduring the onset of terminal alcohol poisoning. I don't think he was even conscious during the cab ride back to the Hotel but once back in his room he awoke, only to speed to the lavatory to be violently sick, and hopefully, to clear his head in time for our next meeting and for the next onslaught of blinis and Vodka Toasts. I was preoccupied with the missile on a stick.

I poured gallons of black coffee into my colleague before marching him out of the hotel and around the block several times to try and clear his head. We endured a more modest Vodka surcharge at a shorter second meeting of the day and returned to our Hotel relieved to have survived. We were greeted there by a Messenger from one of our local Sponsors. We'd been invited to Dinner by the Director of a local Company, and the venue was to be a well-known local restaurant with a Ukrainian 'Folk theme'. I could pretty-well guess what was coming next and my travel companion was suicidal at the prospect.

"Duty calls!" I told him.

We were collected from the hotel that evening by the black combat kit brigade and zipped across the city in the bullet-proof Mercedes, this time to the nominated restaurant where we were greeted at the door. The first thing I saw inside was a life-sized waxwork of a pony-tailed Ukrainian peasant laying on a bed of straw in an agricultural cart, sipping Vodka from a rustic earthenware jug through a straw. My colleague wanted to make a run for it, but I held him back.

The dining tables were long and narrow and of Eastern-European shabby-chic agricultural quality, rather like the battered Refectory tables of my School days, and we sat uncomfortably on long wooden backless benches. Our Hosts were delighted to greet us and Igor, my new friend from the local Company, was quick to pour some good, very good actually, Moldovan Red wine to get the evening and the Party started. Several bottles of this wine were despatched swiftly leading to an interval for some 'Beer Ukrainski' before the wine drinking resumed with the next Course, whatever it was; I'd rather lost touch with what we were eating.

This cycle of wine, beer and food continued until they stopped bringing food, at which point several chilled Vodka bottles were brought to the table and positioned strategically between the diners. Igor continued to smile broadly at me across the Refectory table, as one by one, our dining companions slipped and slumped into Vodka-infused comatose states. One man dropped from the table completely and came to rest on the floor where he slept soundly, snoring and untroubled by any concern from his friends. Igor eyeballed me, as if challenging me to yet another shot of Vodka. Gunslinger time. It was him or me.

The Vodka kept coming and Igor kept smiling, I guess he practised a lot. I'm not a habitual or hardened drinker, but this is where my British Army training really kicked in. Igor might have been a dab-hand with the Vodka, but he'd never had to endure the alcoholic shock-warfare which is the British Army Regimental Mess Night. Unwilling to back down I took him on. In the cold early morning hours, tired waiters carried the semi-conscious Igor back to his car, whilst I dragged my companion back to our Hotel in a Taxi.

The following day and when he finally surfaced, my colleague remembered absolutely nothing of the evening's events, his memory cells had been erased, rinsed away by Vodka. Several years later and quite by chance, I bumped into Igor at the Dubai Air Show. He bear-hugged and kissed me on both cheeks twice, which I took to be a form of homage amongst Eastern European Master Drinkers, and he immediately invited me for a drink back at his Hotel. But I swiftly conjured up an imaginary appointment and declined his kind offer. Yes, I chickened out to save my Liver.

Rosa Klebb's Tomahawk-on-a-stick in the open-air Museum had been one blast from my military past, but another one was waiting for me the following day when I visited a secretive Helicopter Repair Facility. Another old friend was waiting for me there. The Mi 24 'Hind' helicopter gunship emerged from Soviet secrecy during my MoD service, and by the time NATO had classified it I was a serving Infantry Officer in West Germany.

NATO was rightly in awe of the *Hind* in those days as this apparently indestructible combat helicopter was going to pummel to a pulp NATO's forward positions and shoot down anything which got remotely near it. British Rhine Army Planners believed that Russian Special Forces, 'Spetsnaz', would execute daring 'coup de main' operations using *Hinds*, to seize key points and bridges in the vanguard of a Soviet onslaught, and thus *Hind* mythology grew apace; it was said that bullets would bounce off its impenetrable fuselage, etc.

In Afghanistan the USA armed the Mujahedeen with Stingers to take on these fearsome Soviet *Hinds* and in a locally improvised response to the threat, the Mujahedeen shortened the fuses of their RPG-7 Rounds to achieve airburst effects; a technique the Taliban revived when the West succeeded the Soviet Union as killers of Afghans. Afghans have always been good at improvisation and you might have thought the West would have learned that lesson by now.

In Ukraine I was getting up close and personal with these fabled *Hinds* for the first time, having spent a chunk of my military life working out what to do about them. I'd briefed Army personnel on *Hind* capabilities, but I'd never actually seen one fly. We approached the Repair Facility as the first Western visitors to travel there, and several *Hind* wheeled around the sky above us sharing their distinctive acoustic signature. Now I would be an aero-spotting God amongst my ex-RAF work colleagues.

The plinth-mounted *Hind* at the entrance to this secretive Facility had a particularly war-torn ambience about it, and unlike the civilians at the State Aircraft Factory, the Staff I met here were mainly military personnel. I had a long discussion with my genial Officer Host who described how they'd supported *Hinds* in over thirty-five countries. My rust-encrusted memory-cogs clunked as I tried to imagine exactly which thirty-five countries this might have been and I must have telegraphed my thoughts, for my Host smiled and then laughed and said I'd never guess where they all were. He did tell me the story though of the '*Hind*-on-a-stick' at the Factory entrance, but I promised never to reveal the five Countries in which he personally had flown it operationally.

We talked and drank coffee and Vodka of course and I departed late in the day having laid to rest my prior mental image of the invulnerable *Hind*, which nevertheless remains an iconic, and in the right circumstances, a formidable aircraft. After the Fall of the Wall several Western Defence Companies assessed providing avionic upgrades to *Hind* fleets in Eastern Europe and this provoked a huge outburst of Arms Cash chatter and activity.

My final Defence appointment in Ukraine was at a Military Flying Training Base which had once been one of the biggest such bases in the whole of the former Soviet Union. I was dropped at the main gate because civilian cars were barred from entering and after a long and very cold wait I was collected by a General, who appeared through the swirling, freezing mist, driving himself in a big old Soviet Zil saloon. Quite how he managed to get in and out of this car still wearing his huge Soviet-style military hat was a mystery. I was impressed with these big Military hats; they must have had some dual secret tactical function. We have no UK equivalent.

The General was a good-natured soul who apparently spoke no English, so the drive from the main gate to his Office, though peppered with diplomatic smiles, was mutually uninformative. Once inside his Office we were joined by his aide, a sharp young Captain who was fluent in English and various other languages, and off we went to explore the Base.

The General explained that thousands of young Soviet Pilots underwent their military flying training here, becoming proficient in all flight disciplines; Propeller, Fast Jet, Transport and Rotary. The training could last four to five years and the Base had been one of the Soviet Union's premier flight training centres. He was dismissive of Western Air Force 'Taxi pilots' as he called them, with flying skills limited to a single sector of military flying; they were too reliant upon technology he told me. Here, he boasted, pilots *really* learned to fly.

We discussed business concepts as I was interested in the possibility of establishing an international Military Flying Training Centre here on a commercial basis. Looking back now, how ironic was that? There was certainly no shortage of concrete runway or training airspace and the General paused to show me a long flight-line of L29 Trainers drawn up immaculately along the edge of the airfield. He asked me if I would like to buy some of them, all of them even. If I did want a few he added, could I please pay him in aviation fuel as the Air Force didn't have enough fuel to fly these days. It seemed a bit sad for a once powerful Air Force.

Another filed-away memory was stirred by his aircraft sale offer though, as I recalled an Intelligence Report that Saddam Hussein had been planning to convert similar L29/39's into offensive drones, prescient of him in view of the current Ukraine conflict drone activity. Where could those L29/39's come from I wondered? My memory was too vague on the subject so I couldn't really place the details, but in any case, I didn't want any. I'd already been looking at upgraded L59s in the Czech Republic just weeks earlier and had talked to the Czech Air Force about the same possibility for Flight Training. In Ukraine there was both civil and military flight training potential, but it would require significant investment to get this Base upgraded and I doubted that I would find a Corporate Investor. Regional stability, or the lack of it, was a big issue to consider even then, and that was to prove a prophetic thought.

I felt genuinely sorry for the General as it was painful for an Air Force to be grounded for want of fuel. I empathised too, for when I'd served as an Army Operations Officer in Germany, numerous restrictions were placed on Armoured Vehicle 'track miles' and training ammunition quotas were cut. We could neither shoot regularly, nor conduct tactical training properly; fatal economies in an Army. Such basic limitations are demoralising. If a Country wants to boast about having top military Forces, then it must pay for them to train; there are no short-cuts.

Ukrainian Pilots grounded for want of fuel must have been tearing their hair out with frustration as their skills faded, and Simulators, the financially fashionably preferred alternatives so beloved of today's Western Defence Procurement Agencies, are inadequate substitutes for the real thing. An old-world Brigadier I knew used to say there were no substitutes for the 'Three F's': **F**ear, **F**atigue and **F*****ups.

By way of contrast, during my military service with the Oman Forces, I had access to all the training and operational assets any soldier could ever wish for and more besides; tons of ammunition, new weapons, new vehicles, 'Smart' radios, plus helicopters and C130s. It was the end of a golden era though and I'm glad I didn't miss it, as back in the UK Army the comparison was stark. We were just going through the motions in training as we didn't have the resources to do it properly. It verged on being pointless.

When we concluded the airfield tour the General invited me back to his Offices for guess what, yes, Vodka and toasts. A small huddle of senior Air Force Officers assembled for the ritualistic emptying of Vodka bottles in our honour and toasts were proposed and drunk to the new East-West relationship. The chatter flowed as freely as the Vodka until there was a pause, during which we presented our Company 'freebies' to those present. On this trip I had branded document Cases and mini binoculars to give-away. These were of unusually good quality for freebies and were much appreciated. A pregnant pause settled over the group as the General's young Aide nudged me discretely and whispered.

"They are waiting for you to propose your Toast."

Good. I had a plan for that. As we'd progressed around the Base I'd chatted with this young Captain and learned much about his boss, the General, and now it was time to exploit this valuable Intelligence. I stepped forward to address him.

"General, I understand from this fine young Officer…" I gestured towards the Captain "…that in the old 'Cold War' days you were posted for Front Line Squadron duty in what we Westerners then called 'East Germany'… The young Captain translated and the General nodded, looking at me rather quizzically. A hush fell as I continued…

"…and in the event of hostilities, your operational task would doubtless have been to fly ground-attack missions against NATO Forces…" The Captain looked a bit worried and hesitated, but I urged him on. The General's eyes narrowed somewhat, and rather disconcertingly he stepped towards me as I continued.

"...during that same period, I was based across the inner German Border and planned how we might deploy to repel the Eastern Invaders..."

The Aide was very nervous now but this time the General himself urged on the young Officer and took a further pace towards me, glaring hard and muttering.

"...da, da..." to the translation, as if hurrying it along to get to the meat of my little speech. It was time to wrap it up.

"So General, I propose a Toast today to celebrate our meeting here over a glass of Vodka, instead of meeting years ago across a blasted battlefield in Northern Germany... I toast the fact that we never met before and that we are both still breathing..." No pause this time, the General closed the gap between us, laughing aloud, he bear-hugged me before pushing me back at arm's length saying in perfect English.

"We were never going to attack!"

A Vodka-fest ensued during which the General, ever nostalgic for the good old-days no doubt, kept referring longingly to the 'Soviet Air Force' whilst his young Aide corrected him gently. I left the Base with the General's personal gift of a model L39, which for lack of any official reciprocal gift, he removed spontaneously from the young Captain's desk. Today it sits on a shelf in my Study.

I had one day left in Kharkov so decided to spend it looking around the City. I especially wanted to see more Soviet-era sculpture which I admire so much for its energy and scale; I narrowly avoided a second mugging attempt before returning to Kiev, where I sated my military interest by visiting the Museum of the Afghan War. By chance, I found within the Museum a photograph of my new friend, the Air Force General, who was pictured sitting on a rock amidst his fellow young Pilots commanding a *Hind* Squadron in the Afghan Badlands; Western Military Planners really should have visited this place before they lost their way in Afghanistan.

I left Ukraine enthused by the business potential in many sectors and stood in awe of the scale of the country. During WWII German Forces removed trainloads of Ukraine's famous 'Black soil' to Germany, but now it's the USA which seeks rare earth minerals from that same land; so, Ukraine has been geophysically raped as it were, by friend and foe. Because Ukraine's reputation for corruption was widely reported it was no surprise that the country attracted Arms 'Brokers' like bees around a honeypot.

It may have been a poor Nation with a basket of challenges, but I admired the stoicism of the people. Any Nation or population which could endure and overcome a War as vicious and debilitating as 'The Great Patriotic War' must have a core of steel running through it, and an inextinguishable, if well concealed sense of optimism too. Subsequent events have called upon these characteristics to be demonstrated once again.

I'd really looked forward to starting projects in Ukraine and to exploiting my growing contact-base there in Security, Aerospace and Civil sectors, but none of my Defence or Aerospace Principals shared any enthusiasm for this and they all pulled-out after the preliminary work was completed. Now they are rushing to get there. I encountered extensive UK resistance to Eastern Europe generally and few Companies or Governments wanted to know about Ukraine when it really would have made a difference, but they all tried to jump-in once the War started, and in the UK's case, this was mainly because UK Taxpayers were footing the bills.

It was hardly surprising that Arms Cash deals and their commercial equivalents flourished in Ukraine when rhetoric had been the limit of Western interest. The tranquility I experienced on the Ukrainian lake seemed to be as distant and fading a memory, as was my enthusiasm for this work, when I grudgingly took flight to another Eastern European destination, this time to Bucharest. Oh dear!

Chapter 6
Eastern Approaches

Postcards from Bucharest and Craiova

From the moment I landed at Octopeni Airport to the moment I left Bucharest, my first trip to Romania grated in almost every respect, and several subsequent visits weren't much better. I hated the place. My first taxi ride from the airport along the pock-marked road to the city centre Intercontinental Hotel set the tone on my inaugural visit. The taxi, a locally produced Dacia based on a very old Renault design, was an automotive aberration that seemingly self-navigated into every available road crater, whilst the acrid air which gushed in through the uncloseable passenger windows was enough to make one retch. Welcome to Romania.

En route to the city centre I passed a proliferation of the ugliest concrete apartment blocks ever designed and built by man, and I wondered if these places were consciously designed to look this ugly, or if someone had instead misread the construction drawings. These buildings were unfinished and unglazed, their window recesses stuffed with plastic sheeting attached to wooden pallets by the armies of squatters who occupied them. It looked post-apocalyptic; maybe Mad Max lived there.

The Dacia stumbled and farted its way into the hotel drop-zone, and I checked-in. The lobby bar was already awash with hookers and their uniformly red-dyed hair seemed to be the 'must-have' fashion accessory of the day. The hotel was a stubby tower-style building affording some views across the city, although why anyone would want a view across Bucharest was a mystery. I had a high-level room, not of my choice, and one which I'd usually avoid for all the obvious security reasons. The room had a balcony, which I didn't trust from a structural perspective as the concrete walling had crumbled away and the rusty steel skeleton was showing through. Bucharest seemed to be an architectural hell on earth; surely the world's ugliest city after Dubai, so it was beyond my comprehension to discover that during the inter-world-war years, Bucharest enjoyed an architectural reputation as 'Little Paris.' Nicolae Ceausescu and his mates had certainly put paid to that.

To make matters worse, if that was at all possible, I was joined there by an ex-RAF Business Developer, who was so passive that I sometimes felt the need to shake him lest he was actually dead, and the prospect of spending an entire week in his company filled me with dread. I joined him at the hotel bar during my first evening and he was already besieged by a bevy of prostitutes who lacked the demeanour and elegance of their Warsaw or Dubai soul-sisters. Bucharest's ladies of the night descended upon my friend as Jackals do upon a rotting carcass. I navigated through the cordon to join the hapless Executive and shooed them all away. I just hate travelling with others.

I dined with this idiot but left him at the bar after dinner and walked alone to the elevator. A woman followed me and slipped in between the closing doors. It was a well-practised, well-timed manoeuvre. We stood looking at each other. I'd pressed a button to a higher floor than mine and on arrival she left the elevator after me and followed me along the corridor. I stopped abruptly and turned to face her. She was attractive, tall, and assertive, and she pressed herself against me.

"Let's go to your room…" she purred. I stepped back.

"I don't want sex with you…" I said bluntly.

"…I'm very good…" she countered. I didn't doubt it; after all, practice makes perfect.

"Look, it's not for me. Try my friend downstairs…" I said and walked off. She left me to pursue him instead, cursing that she'd still have to pay-off the Concierge who collected money from her and her friends whenever they stepped into the guest's elevator. You must be so careful with these encounters; stay in public areas if possible, don't get into conversations, keep out of touching range and don't be followed to your room.

Work started the next day with appointments in various State buildings and it's usual to take a passport to ensure entry to such buildings. Despite reminders, my travelling companion left his Passport at the hotel, but the office desk staff took his UK gym membership card in lieu of a national ID card, so in we went. I met up there with a Romanian old friend of mine whom I'd first met elsewhere; we'd variously competed and co-operated in other Defence marketplaces before the fall of the wall, when he'd been a State seller of Romanian Defence materiel; geo-political alignments had changed since then. He arranged a mini-programme of visits for us and fixed a trip to an aircraft plant in Craiova.

I was especially conscious on this first visit to Romania to be careful to avoid any agents or Arms Cash conversations as these were a running sore at the time, especially where British companies were concerned. I needed to feel around the edges first, after all, Romania had been virtually a country of informers under the Ceaușescu regime, and at least one major UK-based Defence company, which should have known better, had been caught with their Arms Cash pants down.

In 2003, BAE Systems had been nominated by the UK Government as the prime contractor for the reactivation and upgrade of two old British frigates, *HMS Coventry* and *HMS London*, for the Romanian Navy (yes, there is a navy there, c. 6,500 personnel with six vessels) and this deal constituted the biggest Defence equipment purchase by Romania since the fall of the Communist government. There was a strong political element to the deal as Romania was eager to develop political support for its NATO aspirations, an increasingly common scenario in Eastern Europe at that time.

BAE was 'awarded' this substantial contract but later, UK Serious Fraud Office investigators, together with the UK Ministry of Defence Police, raided the home of Barry George, described then as a 'BAE Agent' and seized documents relating to this deal. Barry George, whom I'd met personally on another Defence matter, was arrested for questioning and then released without charge.

The focus of the ensuing SFO enquiries was a reputed sum of c. £7m worth of alleged Arms Cash, and the story found its way into '*The Diplomat, Bucharest*' and The Guardian in UK, which reported it in detail. It all went quiet until the question of a refund of the Arms Cash to Romania arose in 2016/7. That question brought forward an additional issue when the UK MoD suddenly advised that the contract was 'classified' so the details relating to the request could not be released. As one of my old MoD bosses once said to me, 'The MoD is a cotton-wool empire; the more you struggle, the deeper you sink into the cotton wool and the less you can be heard…' It was not unlike the Israeli government technique for gagging information on arms corruption cases.

Like so many UK Defence exports the deal was enabled via the UK Export Credit Guarantee Scheme, which means that UK taxpayers effectively underwrote the payment of the Arms Cash. So, it doesn't get much worse than that, does it? Well, yes, it does. In the context of rebuilding a new Romania, were these two battered old and hugely expensive frigates really at the top of the national priority list? Who decided on their purchase and why? Or was it simply an Arms Cash-generating mechanism of the type highlighted in the Lord Stokes Report all those years ago? A Romanian admiral subsequently stated that Romania paid three times too much for these old vessels which needed significant refurbishment.

BAE was not the only Defence company in Romania with Arms Cash issues though, as in 2015, the Norwegian Defence group Kongsberg admitted that it had been charged with corruption relating to activities in-country, and in 2014 German Newspaper '*Sueddeutsche Zeitung*' reported on investigations into Airbus Defence and Space Division Contracts in both Saudi Arabia and Romania for Border Systems, allegedly worth around three billion Euros. The Prosecutor's Office in Munich opened investigations into suspicious money transfers amounting to several million Euros, and Airbus Company offices and some Airbus employees' apartments were raided and searched.

Of more immediate interest to me at the time though was a building in Bucharest. As a special treat, my host had arranged a private visit to the 'Palace of the Parliament', that massive State building which had been Ceausescu's pride and joy, although he was shot by his own people before it was completed. The building is truly vast and rates as the second biggest State Building in the world after the Pentagon. My female guide explained that the President personally took a central role in designing the interior, which apparently consumed Romania's entire stock of pink marble. The building has the largest Government Reception rooms in the world and the Ceausescu's planned to hang massive portraits of themselves facing each other at opposite ends of one of these huge rooms, but the unhappy autocrats were hunted down and executed by their own people before these portraits could be completed.

My guide led me out onto the Presidential balcony, positioned as it is to overlook the huge square below, and this balcony too has an ego story. When Ceausescu saw a newspaper picture of himself waving to the assembled masses from this balcony, he was shocked. The huge comparative scale of the building diminished him visually so that he looked insignificant in the subsequent Press pictures, like a doll it was said, and thenceforth he refused to appear there.

The building had been constructed on the iceberg principle with a great mass of the structure underground and many subterranean storeys and galleries. No one seemed to know how many. I asked my State Guide about this part of the building, but she became very defensive on the topic and didn't want to talk about it. She explained that these areas remained out of bounds. Popularly it was claimed that parts of this subterranean structure were used to imprison and torture enemies of the State and apparently some people went in there but never came out again, a la Kremlin.

During the dark Bucharest evenings, I city walked as usual to get more of the flavour of the place. Was there any remaining evidence of 'Little Paris' I wondered? At length, I was delighted to find a few of the older period buildings still standing in between the Socialist concrete monoliths and most of them were now in the Diplomatic area, which was largely closed to the public with access controlled by armed paramilitaries. Bullet-holed buildings dating from the modern Revolution had been preserved for all to see (why do they do that?) but the place quickly ranked highly on my list of places never to return to. Great piles of discarded stinking rubbish lay uncollected in most of the central city areas and the place was teeming with feral dogs and mountains of feral dog shit. They were restoring the Opera House.

The following day I left Bucharest for Craiova in another small car, though not a Dacia, thank goodness, and I sat crushed into the back seat next to my useless colleague. According to the driver, we were going to travel along the 'Highway' the 'Autobahn' the 'Motorway' or some such road, briefly encouraging in me the false belief that there may actually be a decent road somewhere in Romania. This was feckless optimism of course, but less than twenty minutes later, all optimism, feckless or otherwise, evaporated.

The long, long road to Craiova, 'Death Highway' as it was known locally, was in reality a single carriageway carrying an endless stream of horse-drawn Roma Caravans. I thought I'd drifted into a World War II newsreel and become part of a column of fleeing refugees. Pots, pans, and other artefacts dangled from these Caravans and more of the feral-looking dogs and equally feral-looking children accompanied the convoy. Skinny, rib-exposed horses dragged these caravans at a slow ambling, jangling pace. That caravan in the vintage Cadbury's Flake advert would have been a Rolls-Royce amongst these Trailers, and the shapely Flake-biting girl herself would have looked obese here. Where were all these people going to and coming from? There were hundreds of them.

Romanian car drivers delighted in playing their own automotive version of 'Chicken' both with each other and with the Caravan train; the Roma themselves were a disinterested audience. They simply presented a giant slalom challenge to local car drivers with the caravans obscuring any useful views of the road ahead. Not that lack of visibility mattered a jot to these drivers, including ours, as he overtook blindly at every opportunity, until we threatened him with physical violence if he continued.

The sun finally broke through the misery of the early morning mist and damp and heated up the interior of car significantly during this seemingly interminable journey. The two of us, pressed tightly into the small rear seat, almost against the rear windscreen or so it seemed, now felt as if we were melting as the sun's rays were magnified by the rear windscreen.

Eventually, we made it to Craiova, where I'd hoped to refresh myself with a shower in our hotel, but this was not to be. This epitome of Soviet-style Hotels was as welcoming as a kick in the balls and the wooden slatted bed in my room was barely two and a half feet wide, with the type of straw-stuffed palliasse mattress which would have been familiar to Edwardian soldiers. There was no shower and no hot water from the tap. I chose not to use the shaky-looking open-fronted elevator but when I descended via the stairs instead, I had to hunch down because the concrete stairwell had inadequate head clearance for anyone taller than around five feet five. A provincial UK Building Inspector would have condemned the whole horrible structure in a trice, and any Fire Inspector would have done likewise. The hotel food was inedible, so we went out to buy bread and fruit from a nearby market instead of dining in.

We were collected from the hotel the following morning, unrested and unshowered, and driven to the Aircraft Factory. It was a case of *'Kharkov II, The Nightmare Returns'*. At the Aircraft Factory gate-barrier, we paused involuntarily because an angry mob was besieging the offices. Workers were shouting and throwing things at the building. I looked quizzically at my Company host who shrugged.

"Annual wages negotiation…" he said casually. "…we go around the back…"

Once inside the Factory I enjoyed a tour of the manufacturing plant which was set up to assemble the indigenous military jet training aircraft, the IAR 99, called the '*Shoim*', which was described to me as a Romanian version of the UK Hawk Jet Trainer. Apparently, *Shoim* translates directly as *Hawk*. Especially interesting was the installation of an advanced digitised 'Glass cockpit' into the IAR 99. This was more advanced at the time, than any equivalent RAF training jets, courtesy of my old friends Elbit Systems from Israel. Elbit had its hands full elsewhere in Eastern Europe, though, with Arms Cash allegations in Bulgaria, but here in Craiova, there were no apparent issues.

There was some UK content in the IAR99 insofar as the engine was the old Rolls-Royce '*Viper*' model, noisy, but potent enough to give the plane a robust performance envelope, as demonstrated at Air Shows by a highly capable Israeli Test Pilot whom I met later in Tel Aviv. I made a presentation to the Base Commander and an audience of Staff Officers about how to provide Military aviation training facilities using Contractors instead of Service personnel, and using Private Finance Initiatives (PFI) in place of capital spending from the Air Force budget to provide the aircraft. They listened attentively and discussed the proposition between themselves there and then. It was well received. The proposition was straightforward and the Romanians we met wanted to do it and Elbit seemed to want to join in too, so the concept could have worked well for the Romanian MoD. The pioneer project had been set-up in the UK.

But my Company flunked it yet again with no appetite to pursue this business, so having once again excited everyone else about the potential I had to pull back; no deal. It was left to another Company to gleefully pick up the concepts and profit from them using the same model. It was another depressingly predictable and familiar outcome and other UK Companies also showed little interest or ability to follow up or exploit other potential projects I found in Romania. If a business proposition was innovative, creative and south of Dover, it would be strangled at birth. I don't know why the Companies I've represented over many years have pretended to have overseas ambitions as they never pursue any new international prospects, even when face-to-face with the Users who want them to do so, and with budget or finance available. A mystery. Some Projects are born to fail whilst others have failure thrust upon them, and failure was certainly thrust upon my Romanian prospect.

On later visits to Romania with a different aerospace company, their Directors flunked another big opportunity, this time to purchase an aircraft repair facility with massive regional and broader potential. Whatever happened to the spirit of enterprise? The entrepreneurial aviation pioneers of the 'twenties and 'thirties who founded the UK Aerospace Industry would have turn in their graves with dismay and disbelief; many and frequent were the examples of commercial cowardice and ineptitude I encountered. Today's Companies are not run by entrepreneurs, they are run by safety-first Corporate politicians and Risk-aversion addicts. Got the message yet?

There was one entertaining aviation footnote to my visit to Craiova though. When we'd finished our meeting in the Airbase, the Air Force General asked me how I'd travelled to Craiova, and not wishing to be dishonest I told him I'd survived the harrowing car journey along 'Death Highway'. He chuckled at the thought of my discomfort and invited me to return to the Capital with him, although I didn't know how the General usually made this journey. However he did it, it was likely to be more comfortable than my experience of dodging Roma caravans and feral dogs so I accepted his offer blind, surely, it could be no worse than our outward drive?

I fell-in behind the General and his entourage as he marched with commendable briskness to the nearby airfield, where waiting for him was another of the twin-engined Antonovs which I had grown to love so much in the Ukraine. This particular airframe was rigged for military use and the General gestured to me to follow him as he walked up the aircraft's rear loading ramp and into the cargo/paratroop area.

Plastic-covered bench seats were fixed between the fuselage ribs, but there were no belts, webbing, buckles or slings; no visible restraining mechanisms of any sort. As he walked up the ramp the General took off his huge General's hat and tossed it casually over his head and an adroitly positioned Aide deftly caught the spinning hat and tucked it under his own arm; a quietly impressive fielding performance. Clearly the General was a man who knew how to handle not only his Staff, but also his hat. Now, that's impressive.

Once aboard the aircraft, the General sat down and swung his feet up onto the bench seat and lay back, resting his head on a rolled-up blanket which was positioned there just in time by his hat-catching Aide. The General closed his eyes to sleep and someone gave me a can of Orange Tango. I don't know why. I grabbed hold of a fuselage rib as the loading ramp was raised and the plane started to taxi. It was a shuddering, juddering take-off and an equally shuddering, juddering low-altitude bumpy flight back to Bucharest, but for all that it was infinitely better than risking life and limb again on Death Highway. The Antonov stayed very low all the way back to Bucharest and I spilled most of the Tango onto the plastic seat, but eventually we landed at a military Airbase and the General instructed his Aide to call a taxi for us. It was a welcome gesture. He bade us farewell and left in his Staff car.

I've been in and out of Bucharest several times subsequently, although Romania had not improved much on those visits; I understand it's better now. In fairness I can say two things in Romania's defence. Firstly, I found a second-hand bookshop in Bucharest in which I bought a rare and interesting aviation book (you'll just have to trust me, there is such a thing) written in English and which I'd never seen subsequently anywhere else. Secondly, I discovered the work of Romanian artist Nicolai Grigorescu, which I enjoy very much and which I was glad to discover. So, all was not lost. Who knows, after another fifteen or twenty visits I might just find something else about the place to enjoy.

The later Russian activities in Ukraine and Crimea prompted Romania to address its national underinvestment in Defence equipment and in 2019 it was reported that the Country had placed significant contracts to upgrade those two expensive ex-UK Frigates, now re-named *Regina Maria* and *Regele Ferdinand*, though not with UK Companies, and also to enhance their national Command and Intelligence capabilities. Whatever I may have gained or lost personally in visiting Romania, I must have convinced someone at home that the Eastern European military aerospace market was worth a shot, as I was soon despatched again into Eastern Europe, this time to Warsaw.

Postcards from Warsaw and Kielce

Poland joined NATO in 1999 and was one of the first ex-Warsaw Pact countries to achieve membership along with Hungary and the Czech Republic. Poland's NATO accession made it an instant target for

Western Defence Companies which had been novelties in Poland up to that point. Sadly, the co-operative industrial and technological hopes of many Polish Companies were cynically and artificially buoyed-up by smooth-talking Western Executives who smooched with the Poles and hinted at all sorts of two-way trading possibilities with Polish Companies.

Of course, this early commercial flirting was mostly bluff, since Western companies just wanted to sell their own products and systems into the country or get stuff manufactured in Poland on the cheap. The marketing premise was a simple one; Polish Forces would need to equip upwardly to become interoperable with her new NATO buddies and would have to buy mountains of high-technology Defence equipment to make this possible. All Western companies had to do therefore, was to turn up and harvest the money-tree. It was a familiar marketing script lacking reality or data, but it was enough to get us all on planes to Warsaw and off we all went.

Similar scenarios played out all over Eastern Europe as the Soviet shadow receded and Defence Industry cynics observed that NATO had expanded simply to create bigger and additional markets for Western Defence Contractors, especially those from the USA. It became a depressingly familiar commercial exploitation story with disappointments on both sides, but mainly on the Eastern European side. Ironically, Russian aggression in Ukraine has now boosted Western Defence Company revenue by a huge margin, so the NATO strategy of pushing provocatively further and further Eastwards has certainly worked-out for Western Defence Companies.

Poland was a prime early target Customer Country owing to its scale, accessibility, and a viable-ish economy, and there were historic links too, certainly with the UK, which did after all go to War for Poland in 1939. But the simple geopolitical reality was that Poland was regarded by the West, i.e. the USA, as a key Buffer State which needed to be equipped well enough to buy time for the rest of the West in the event of Russian aggression.

From the American perspective, Europe and Eastern Europe in particular, appeared to be simply US weapons and Systems platforms in a strategy of keeping any future war away from mainland USA, in which respect Eastern Europe became a sort of American Aircraft Super-Carrier. The constant eastward creep by NATO despite assurances to the contrary, was only ever going to fuel Russia's security concerns, but no one was paying attention. Now look where we are.

I visited the Warsaw-based Aerospace Company PZL, a Company with deep roots in the earliest and arguably the most exciting years of the Aviation Industry, the late 1920s and early 1930s. PZL had a long heritage of aircraft design and manufacture, even though their Plant was all but destroyed during the German retreat in World War II. Soviet annexation of parts of Poland and their subsequent political domination of the Nation after the War, brought the capacity of PZL into the Warsaw Pact's industrial portfolio and thousands of PZL aviation workers were sustained in the Soviet-style 'Command-Economy' of the Cold War era. When the Wall fell it was a mixed blessing for PZL.

Poland changed politically and economically afterwards but the fortunes of PZL plummeted. The Company no longer benefited from guaranteed State orders to keep it busy but instead had to fend for itself in an alien and competitive international market, in which it was fatally handicapped by the triple burdens of an oversized workforce, inadequate finance and outdated technology. The Company was forced into a commercial tailspin from which the only flimsy recovery path was release from State ownership. As a final insult, PZL was sold to that paragon of Euro-homogenization, Airbus, and PZL's joy on the advent of this new ownership was summarised for me by one senior PZL Executive thus '...*since the 1920s we'd designed and built countless aircraft, but as part of Airbus all we were trusted with was manufacturing Cargo doors...*'

PZL had indeed designed generations of Polish aircraft but was now just a piece-part manufacturer in the great pan-European business model. The Company which had responded to the commands of a Communist regime now responded to the commands of a pan-Euro corporate monolith. I'm not sure they could tell the difference; I certainly couldn't. There was a subsequent note of optimism though. With typical Polish aeronautical technical excitement, stoicism and determination, a small PZL team made the best of their new situation by upgrading their existing Polish military Trainer and they digitised the aircraft's Instruments by installing a Glass digital cockpit, just as the Romanians did in the IAR99. Poland and Romania were thus both years ahead of the UK RAF's vintage basic training aircraft, the ageing and barely fit-for-purpose Tucano, the replacement for which had been the subject of expensive serial dithering by the MoD. The Tucano had no digitised cockpit despite the need to feed trainee RAF Pilots into highly digitised Fighter aircraft. It may have been small beer commercially, but PZL demonstrated the Company's ability to adapt and survive, lessons lost on many British Companies which either sank without trace or were bought by overseas buyers.

The twin seductive sirens of the EU and NATO membership had promised a post-Cold War land of milk, honey and money, but few Eastern European States anticipated the commercially subservient reality which ensued. Indigenous manufacturers were courted by the West with the promise of industrial participation and new programmes but were left disappointed. It was more truly a form of industrial suppression. Heavily prejudiced Western instincts against virtually anything of Eastern European origin killed off most projects and no one was buying seriously from the East, just selling.

I could garner no support for or interest in, any project involving Eastern European aircraft or their Manufacturers, despite their unquestionable technical heritage and the existence of specific market opportunities. It confirmed yet again that when it came to co-operative projects, that infamous modern Defence Company risk-aversion philosophy kicked-in and coupled with an immovable 'not invented here' syndrome, represented the kiss of death for most Eastern European opportunities. Western companies wanted to play in this regional Market but only if they were guaranteed leading, controlling roles and a win before they even started. The Israelis, attentive as ever, were left to clean up in niches where these Defence and Aerospace companies feared to tread, as in Romania.

A subsequent Polish National response to the new and evolving Defence market landscape was to create a composite National Defence Sales Organisation of their own, called the *Polish Armaments Group (PGZ)*, which came into being in 2013. On foundation this Group combined over sixty different Defence business units and employed over seventeen thousand people. But events in 2016 exposed Arms Cash investigations by the Polish Central Anticorruption Bureau (CBA) and the Bureau announced that various individuals had been detained, including M. Bartłomiej, a former head of the political cabinet at the Polish Ministry of Defence, Mr. K. Mariusz Antoni, a former MoD official, and Mr. Radosław, a former PGZ Board member, plus various Polish MoD Officials. As with many Arms Cash cases, these events were kept under wraps until a newspaper, 'Sieci' broke the story, stating that:

'A business-friendly system was created around one of the largest state-owned companies to siphon public money using ostensible contracts, fictitious training and inflated invoices. The network reached the level of PGZ's management and the political cabinet of the Ministry of National Defence

The Polish Defence Ministry denied these allegations initially but in 2019 Radio Poland reported that six people had been arrested and that investigations were ongoing. Two years before the investigations started, it was Hewlett Packard who'd been in trouble when it was reported that they had been charged by the US SEC with violating the US Foreign Corrupt Practices Act (FCPA) in Poland. It was asserted that HP Executives *'provided gifts and cash bribes worth more than $600,000 to a Polish government official to obtain contracts with the national Police agency.'* H'mm, chicken feed in the great Arms Cash scale of things.

I visited Warsaw on another occasion to promote high-tech military comms equipment for a US Defence Principal, this time working through a local Company in a Distributor mode, as mandated by the Polish Government because of the technical (i.e. secret) nature of the project. Our equipment was massively over-priced and over-specified for the task, but the Poles had been convinced, or more probably had been told by NATO that the equipment was essential for NATO interoperability, so we had a clear chance at winning a contract. We had a tough time getting through various technical obstacles with Officials from the Intelligence Staff but eventually made progress.

To celebrate this little breakthrough, I was invited to join the MD of the Polish Distributor Company and his ex-Polish Army Colonel Manager for dinner in the Old Town. The surviving parts of Warsaw's Old Town are sparse because the Nazis smashed up most of the City after the Ghetto Uprising, adding to the extensive damage already wrought by the Luftwaffe, but what remains of the original, especially the Market Square area, is charming and was undergoing refurbishment and restoration work at the time of my visit. Elsewhere in Warsaw, numerous modern buildings of questionable architectural merit have sprung up to replace the previous Communist-era constructions. Some of these architectural fantasies are arguably more painful eyesores than were their smashed-up predecessors, which qualifies many of them as candidates for the Dubai cityscape.

Dinner dish of the day was the national dish of every day, Golonka, an enormous pork knuckle served up on a folksy wooden platter; a roughly sawn plank surely, along with a garden fork and a small hand saw with which to attack this culinary beast. Food was washed down with copious quantities of Polish beer, very drinkable, with intervening shots of 'Mad Dog,' a locally popular down-in-one cocktail combination of Vodka, blackcurrant cordial and Paprika. An additional culinary treat came in the form of dollops of pork lard, dense and white, self-served from a communal wooden bowl in the centre of the table. I apologise for forgetting the local name of this delicacy which I hope never to see or taste again. It was a digestive challenge of Olympic proportions, with successive mouthfuls of bullet-hard crunchy pork-crackling breaking my teeth, and artery-clogging dollops of the dense white lard. Guzzles of Polish beer and 'Mad Dog' shots provided liquid interventions.

I really, really should be compensated for these dietary trials on official business, especially for the simply revolting spiny Hungarian Pike dish with which I once battled in Budapest, and which wounded me at every bite; it too was a local 'delicacy.' My Oman military service inured me to the challenges of traditional Middle Eastern cuisine, which seldom appears anyway in the region these days owing to the spread of diner-friendly Western menus. I doubt that many young people in the Gulf these days have ever actually eaten a goat's head, although I know for sure that I have.

The evening wore on and the 'Mad Dogs' flowed freely until I reached a point on the alcoholic spectrum where I knew my grip on reality was slipping away. I said a polite goodbye whilst I still could and left the party, opting for a brisk long walk back to my hotel through the cold night air to counter the alcoholic excess. The Polish ex-Colonel business manager of the local Company, his bright blue eyes shining through the alcoholic mist, grabbed me as I was leaving and kissed me goodnight on both cheeks, lingering a little too long in the suction position for comfort. A slightly awkward moment. Similarly, as is the Arab custom, Officers of my Omani Regiment held each other's hands in friendship and sometimes held mine too. It was neither something to recoil from, nor something to enjoy, it was just

176

their way. When in Rome...! For now, I was just happy to be out of the restaurant and city-walking on my own again.

I found a great viewpoint on an elevated terrace overlooking the Vistula and on that crisp starry night, the great river in full flow was a magnificent sight, with the majestic arcs of Poniatowski Bridge linking the banks. The bridge had been damaged badly in both World Wars and rebuilt twice, so in its way, it symbolises the resilience of the Poles. Polish bridge-building Engineers had done good work in Warsaw, much as the bridge-builders had done in the other great Eastern European Cities which I was now visiting. I was collecting bridges. The Vistula bridges carried invading German Forces into the city from the West, whilst the Soviets later gobbled up Poland and crossed the bridges from the East. The Poles seemed to attract Invaders, so history owes them a break; they certainly weren't going to get one from any Western Defence Company though, that was for certain.

The modern Polish Republic was founded only in 1918, so Poles have had more than their fair share of invasions and uprisings since then, but they stand aside from most countries because they had the good grace to install a famous Pianist, Jan Paderewski, as their first Prime Minister. Music and Poland sit well together.

Back at my hotel, I drank several cups of coffee and watched the ladies of the night stalking their prey. These Polish girls were noticeably more smartly dressed than their Romanian counterparts and I was propositioned as I headed for the elevator, but declined as usual and kept walking, claiming not to have any money. The price dropped immediately and as I entered the elevator, it dropped again. The doors closed and thankfully separated us. I slept soundly thanks to the 'Mad Dogs.'

It was real dogs not 'Mad Dog' cocktails, which fed from the kitchen debris at the Polish Defence Exhibition Centre at Kielce, which I visited on another trip to Poland. The highlight of that visit had been the Fine I'd paid for taking a tram ride without a ticket, and that Fine remains the dominant memory of the trip, which says something about my level of interest in the Show itself. It was an old-fashioned and depressing show because the exhibits were predominantly Warsaw-Pact derivative tanks, APCs and aircraft. It was mostly hardware, and old military hardware at that, which dominated over technology on display.

By way of contrast, Western Defence Shows had gradually transitioned to soft-core digitised offerings such as ruggedised computers, C4I systems, Simulators and of course Drones, and the old Soviet 'T' series tanks displayed at Kielce had predictably been over-matched by Western equipment during the Gulf conflicts, so had lost their appeal. Curiously, the Afghanistan intervention (invasion) forced the Western Defence Exhibition pendulum back the other way for a while, so that actual guns reappeared in profusion at shows once more, and there was a sudden concentrated focus on Mine-resistant and armoured vehicles. The marketing whims of Defence Industry Executives and military necessities seldom align. Today everyone loves Drones but who knows what it will be tomorrow? My money is on Drone-mounted ethno-selective targeting sensors. Watch this space.

Eastern European Defence Procurers take note though. One notable outcome of the failed Western Afghan invasion was that despite buying and deploying billions and billions of dollars-worth of high-tech military equipment and enduring (by Western standards) major loss of life, Western Defence high-technology failed to suppress the insultingly-labelled 'stone-age' Taliban enemy. The world's most powerful military Nation was humbled and finally fled Kabul like a thief in the night. Nobody told the Taliban that all this advanced Western technology would result in their defeat; they didn't or couldn't read the Western script. In Churchill's words '*No matter how enmeshed a commander is in the elaboration of his own thoughts, it is sometimes necessary to take the enemy into account*'. USA, Britain and their Allies clearly did not.

Travelling through Eastern European States in the aftermath of the Fall of the Wall also gave me a chance to see what Wellington described as '*...the other side of the hill*'. Most of these States hadn't seen the huge NATO costs coming down the line when they applied for Membership and were only just starting to grasp the economic implications and reality of it all. They all wanted Uncle Sam's protective arm(s) around their shoulders but hadn't planned on paying-over massive chunks of their GDP for the privilege. Now of course Uncle Sam seems to be packing his bags and taking flight. In my world I prioritised these States according to a variety of factors and depending on which Company I was working for at the time. I did not prioritise Bulgaria though. Someone sent me there against my will.

Postcard from Sofia

If Poland had been at the top of the Defence Company Eastern European Prospects list then Bulgaria had certainly been at, or close to the bottom of it. Nevertheless, I found myself in Sofia, ugly city, and soon discovered that the Bulgarians too had Arms Cash challenges.

Defence Minister Nikolay Mladenov stated that Bulgaria had probably suffered losses of around 100m Bulgarian Lev from several national Defence deals and confirmed his intention of changing Bulgarian Defence laws to eradicate such occurrences in the future. Former Defence Minister Nikolai Tsonev was variously charged, tried, and acquitted three times in connection with Defence corruption. Here was another classic example of an 'Influencer' at work.

State Prosecutors alleged that Tsonev had signed off four questionable deals in 1999 which had cost the struggling Country nearly five hundred thousand Sterling (equivalent) and they demanded a multi-year jail sentence for him. Tsonev was acquitted of charges on another count that he had been involved in a deal which had led to 'material damages' for the Ministry of Defence running as high as 13 million Levs. Tsonev was not alone in being connected with Arms Cash allegations in Bulgaria for two other Senior Officials were also found guilty, although they were allowed to remain free pending appeals.

My old friends Elbit from Israel featured once again, when Bulgarian newspapers *Douma* and *Monitor* claimed that the Company was involved in the scandal with the Bulgarian MoD. When Tsonev was arrested it was reportedly in connection with the contract Elbit won to upgrade helicopters to NATO standards. *'We are talking about high-level corruption. Millions of levs are in question,'* said Jordan Bakalov, Chairman of the Bulgarian Parliament's Citizens' Complaints and Petitions Committee. He could have telephoned the UK MoD to ask how they had coped with the discovery of Gordon Foxley's Arms Cash deal in the UK.

I made brief visits to Sofia with Tactical Comms' equipment, though a competitor had visited before me. Fresh and green from military service, that Company's Business Developer made undeliverable Offset promises to the Customer with no real commercial understanding as to how Offset works, and his moment of madness was my opportunity. There are numerous well-meaning but useless ex-military types in Defence Sales roles for which they have no appropriate skills. It is not enough simply to

179

have served in the Military to succeed in Defence Sales and in my experience the reverse is clearly more usually true.

I was interviewed once by an ex-Colonel Sales Director whose entire Sales training experience consisted of an Army Resettlement Course during which he'd role-played selling Pizza. He revealed this to me in confidence as he didn't want his Sales team to know how limited his Sales expertise really was, even though the lack of it shone forth on a daily basis. He simply didn't 'talk the talk' and seemed to be the only one on the payroll not to realise it.

I left my competitor to it in Sofia after floating a teaming proposal with him, but his Company had no 'receive mode,' so it was pointless. He'd raised Client expectations unrealistically and wasted a really good Agent who simply couldn't understand what he'd been playing at with such a bizarre proposition. His US Parent Company sent over an American Minder to watch and report on progress, so there were two different report lines running back into the same Company; which is not unusual. It was the perfect Defence sales f***-up; a faulty proposition, a second-choice system, an unbriefed Agent, and a US Minder reporting back on every move. It worked out well for my company in the long run though, but the Bulgarian MoD had more pressing problems.

In 2021, the Bulgarian ammunition Company VMZ-Sopot was revealed by a Parliamentary Audit Committee as having transferred $10m to a Company in Serbia without receiving anything in return for the money. This payment was supposedly part of a larger $20m deal, but official investigations were hampered because the Company Board claimed that they didn't have '...*access to the documentation*...' as the deal was '...*classified*...' (that old chestnut again). It was suggested that the $10m had been paid as a '*Fee.*' Whatever could that be I wonder?

Several other allegations against Bulgarian Defence Companies were being investigated at the same time and curiously, Bulgaria declined to participate in an EU ammunition-supply program for Ukraine. Later, however, the 'UNIAN' News Agency reported that Bulgaria would transfer large quantities of ammunition to Ukraine but only '*through Intermediaries.*' I'm sure this had nothing to do with Arms Cash. Corruption within a State can attract international sanctions and have broad political ramifications, as evidenced by this excerpt from a US State Department Press release dated 10th February 2023:

'...The United States, in coordination with the United Kingdom, is taking action to counter systemic corruption in Bulgaria by designating five former Bulgarian government officials as well as five entities for corrupt acts that resulted in illicit personal gain, undermined the country's democratic institutions, and perpetuated its corrosive dependence on Russian energy sources....'

This Press Release, issued under Antony Blinken's name, goes on to name Bulgarian individuals sanctioned by both the USA and UK for corruption, but readers could be forgiven for spotting the glaring irony here. Whilst the UK was quick to align itself with the USA in condemning a few Bulgarians for their actions in Bulgaria, UK authorities had been unwilling to investigate major British-based Defence Companies for their own multiple Arms Cash activities.

I didn't dwell long enough to city-walk in Sofia so apart from trudging the length of the main street to and from the Intercontinental Hotel, I saw little of the city. Mostly I saw the inside of offices and my hotel room. Not unusual in my world. Nor did I see anything of the surrounding countryside, although I was assured that excellent and cheap skiing was available in the nearby mountains, but I don't ski so this doesn't matter to me. Bulgaria became a pretty duff experience, especially when compared with the joys of the Czech Republic. I'd always wanted to go to Prague, so when the chance arose, I jumped at it.

Postcards from Prague, Brno and Zlin

The Czech Republic was the preferred destination for most Western Suppliers during the initial Fall-of-the-Wall Defence Company goldrush into Eastern Europe. Western Companies were quick to display their wares at Defence Exhibitions in Brno, the town which gave half of its name to the famous *'Bren'* gun. Like Poland, the Czech Republic was relatively stable and after the velvet divorce from Slovakia everyone wanted to visit Prague again, that ancient city of a thousand spires and the centre of so much Cold War intrigue.

Who could fail to enjoy a city-walk in Prague, especially across those famous bridges (yes, more bridges) spanning the Vltava, or take a leisurely wander through the Old Town Square, pausing to gaze in wonder at the ancient Astronomical Clock? In my old UK MoD days, the former Czechoslovakia was yet another place I wasn't allowed to visit, at least not without completing a complicated Approval Form and having to justify why I wanted to go there. Since then I've been in and out of the Czech Republic on business many times along with millions of tourists.

The first time I travelled to Prague was on military aerospace business, but I later visited to discuss Command & Control Systems and various other things, including SF training facilities. I had good senior contacts in Prague, both military and Industrial, and there were good co-operative opportunities with Czech Aerospace and Defence Companies. But extra care became essential in the Aerospace sector because of Arms Cash issues.

A BAE Systems-SAAB Consortium planned to sell Gripen jets to the Czech Government in a deal which included significant Offset work in the package, but after much wheeling and dealing the Czech Parliament declined to support the expensive proposal. The result was a modified proposition worth a reported £400m, in which the lesser number of c.14 aircraft would be leased instead of purchased outright.

Czech Investigators probed the possible Arms Cash aspects of the deal and subsequent allegations swirled around the deliciously named Austrian Count Alfons Mensdorff-Pouilly, who allegedly operated as a Middleman or Agent. Investigators were stalled by a lack of co-operation from both the countries involved, not least by the UK, but the situation was supposed to be improved when the British Prime Minister allegedly intervened to promise UK co-operation with Investigators.

It was later widely reported that the UK Serious Fraud Squad interviewed the Count about the Deal and charged him in 2010 with conspiracy to corrupt in connection with BAE's deals with various eastern and central European countries, including the Czech Republic, Hungary and Austria. The Count was also arrested by Austrian authorities and questioned about BAE's deals and the payments made to him worth an alleged £11m. At his Vienna Trial in 2013 the Count claimed that he was merely an 'Adviser' to BAE, and he received a two-month suspended sentence for 'falsifying evidence'. Judge Stefan Apostol told the Count *'The whole thing stinks.'*

As is usual in such cases it was alleged that Arms Cash was channelled to various Officials via dark Accounts in discreet banking locations but this was never proved in Court. UK Investigations came to a halt when a controversial all-embracing Plea Bargain settlement was announced in which BAE pleaded guilty only to 'Accounting irregularities' over its 1999 sale of Radar to Tanzania. The Company paid a UK Fine of £30m in return for which the SFO did not pursue the other corruption allegations, including those relating to Czech Republic, South Africa and Romania. BAE did pay a world record-breaking Fine of $400m in the USA though, and a US Department of Justice statement included the following passages:

'BAES was sentenced today by U.S. District Court Judge John D. Bates to pay a $400 million criminal fine, one of the largest criminal fines in the history of DOJ's ongoing effort to combat overseas corruption in international business and enforce U.S. export control laws. BAES admitted that, as part of the conspiracy, it knowingly and wilfully failed to identify commissions paid to third parties for assistance in soliciting, promoting or otherwise securing sales of defence items in violation of the AECA and ITAR. BAES failed to identify the commission payments paid through the BVI (British Virgin Islands) entity described above, in order to keep the fact and scope of its external advisors from public scrutiny. In one specific instance, BAES caused the filing of false applications for export licenses for Gripen fighter jets to the Czech Republic and Hungary by failing to tell the export license applicant or the State Department of £19 million BAES paid to an intermediary with the high probability that it would be used to influence that tender process to favor BAES.'

In the separate case of the EADS (now Airbus) Eurofighter sale to Austria, the total Arms Cash package was calculated at Euro 120m and the 2017 Austrian MoD investigation into this deal resulted in a Civil lawsuit with damages calculated at Euro 1.1bn, with the CEO of EADS being named as one of sixteen 'people of interest' in the case. Here was another classic example of Influencers at work and the German investigation into Airbus was terminated with a Settlement Fine of a staggering EUR 81.25 million.

The general Arms Cash situation in the Czech Republic wasn't improved by a March 2011 Report that revealed that Investigators had uncovered large scale activities involving some thirty Czech Defence Tenders and that those under investigation included eleven ex Defence Ministry Staff members and nearly forty Company Executives or Managers. The Tenders thus tainted were valued at c.300m Czech Crowns. In February 2011, anti-corruption Police indicated that formal charges should be brought against fifty-four individuals, making this one of the largest known Arms Cash cases in recent European history. The criminal files apparently consist of nearly twenty thousand pages of material.

Later developments followed-on from another major Czech Arms Cash scandal in which the Deputy Minister Jaroslav Kopřiva was removed for attempting to set-up an Arms Cash deal for vehicle-mounted Mortars manufactured by Finnish company Patria. In this case it was planned to bypass the procurement Tender process by arranging what appeared to be a joint procurement deal with neighbouring Slovakia and Minister Kopřiva had even lined-up his own successor to ensure continuity in the deal if he lost his own post.

My humble 'through the front door' aerospace marketing in the Czech Republic paled into innocent insignificance in comparison with all this high-level Arms Cash activity which provided excellent examples of both 'Influencers' and Agents at work. My early interest in the Czech Republic included the Zlin light aircraft which could be a candidate for the various overseas military flying training Offers I was working on, and I discussed these possibilities with the manufacturers.

Overseas Air Forces had seen the UK RAF adopt a highly controversial 'Private Finance Initiative' (PFI) approach to basic Flying Training in which contractors, not the Air Force, provided aircraft and Instructors, and curiosity ran high as to how this might work in the Czech Republic. Combining a proven Czech light-aircraft with the commercial clout and international reach of a UK-based Aerospace Services Company should have been a dream made in heaven.

I'd assessed several related Middle Eastern Air Force opportunities where such a deal could work and especially at one in Egypt. I used to meet Hector's Lawyer Essam in Cairo, so it was nice to get back there and wallow in a little Arms Cash nostalgia. The robust, highly capable Zlin would be a highly effective and affordable candidate for any basic flight training task, although aviation snobbery kept it and comparable aircraft out of the RAF's procurement competition.

The RAF's own light aircraft requirement was interesting because the contracting mechanism was a complex Contractor-managed PFI flying-hours deal in place of outright purchase, so the RAF was not incentivised to be cost-efficient in their choice of airframe. It was as if a Private Finance deal made it possible for the RAF to have anything they wanted, and their instincts thus inclined them to the most expensive Training aircraft they could get their hands on as their preferred airframe.

They selected a new *Grob* light aircraft from Germany, which had not even entered serial production at that time. Grob's core business had been Gliders. The proven world beating Zlin aircraft didn't even make the cut but neither did UK's indigenously produced light aircraft, manufactured by Slingsby. It's hard to imagine any other Country rejecting its indigenously manufactured aircraft in similar circumstances. No licensed manufacture of the Grob was agreed so the whole fleet of c.100+ aircraft was manufactured in Germany with no reciprocal manufacturing or value-add for UK Companies, and questions were asked in the House of Commons about the programme.

I visited the Czech town of Zlin with a former RAF Officer to see and fly in a Zlin aircraft, but because the UK MoD had a cultural block on anything built in Eastern Europe it was pointless; a nasty case of product apartheid. The world-aerobatic champion Zlin aircraft, proven both in the air and in serial production at around half the price of the unproven Grob aircraft, never got a look-in for the RAF Requirement. Grob Aerospace filed for insolvency in 2008. A new aircraft could have been purchased and manufactured in UK and created jobs and a downstream market for a national Manufacturer, but this option was not even considered. Eyebrows were raised amongst the knowledgeable.

I've marketed contractorised Defence Services to Forces around the world with mixed outcomes as not all Armed Forces share UK's enthusiasm for handing over Military tasks to private Contractors. Equally, some indigenous Service providers have been quick to spot the Arms Cash potential in this style of deal. Some overseas senior Officers I met who wanted to try Private Finance Initiatives' (PFIs) or other contractorised Services proposals, were restrained by more conservative Defence Ministries which said no. One Air Force Commander I met even set out to get national law changed so he *could* contract on a PFI basis and therefore get the Training aircraft he needed but could otherwise not afford. In another country a Minister asked me to write a PFI Brief for him to present at Cabinet to pave the way for his MoD to engage with the Company for which I was working. A Ukrainian City Mayor asked me if I could originate a City mass transportation PFI project.

Working in Czech Republic had been fun, but frustrating from the UK side of the equation with all the usual resistance to new export business at home. Risk aversion had become a religion by then and no doubt the high-level Arms Cash incidents in the Region put off some Companies. Prague had been super and remains a favourite place to visit, but it was time to move across Europe to another country I'd longed to visit, Hungary. Maybe they would 'buy into' PFI concepts.

Postcard from Budapest

On my first visit to Budapest the Hungarians were not too bothered about any of this PFI or contractorising nonsense, as they were busy hosting visits by Western C4I Systems companies (*Command, Control, Communications, Computers and Intelligence*) which were flavour of the month amongst NATO new boys. I'd previously worked on a C4I System proposal for Hungary using an adaptation of Prof. Kahn's work on Conflict Escalation as the basis for the System's architecture, to 'read' social and political escalation events as triggers for layered responses, civil and military. It would be a gem for AI today.

Now I was back in Budapest representing an Aerospace Company and pitching the benefits of contractorised and PFI services. An amiable Air Force General paused our meeting to usher me outside, where he gestured towards the large number of young Air Force personnel hard at work sweeping leaves, painting fences and suchlike. He asked what would I do with all his personnel under a new contractorised system? Would they all get jobs or not? It was a fair question. Probably not was the answer.

Manpower economy might have been a hot button at home, but it wasn't so in Hungary at that time, in fact it was more of a threat. 'Ownership' of capital military assets was another cultural issue. Another Eastern European General I met could not break from the ethos and habit of ownership and he would only contract if the Air Force, not the Contractor, owned, controlled and serviced the aircraft. I'd have to go back there a few years later with such a contractorisation concept. I did write a PFI Brief for his Minister though, I should have charged for that.

I stayed in Budapest in yet another Intercontinental Hotel, but no, I don't have an Intercontinental 'habit' and this particular Intercon enjoyed a great position close to the Danube. It was late Autumn and a wonderful time to be in the City, with melting sunsets and the great

186

historic artery of a river in full view from my room. Sadly, not the Blue Danube of musical fame though, it was a muddy, murky brown Danube and surprisingly fast flowing for such a broad river.

The great Chain bridge which links Buda with Pest was, I was reliably informed, designed by the same Engineer who designed the much smaller version which spans the River Thames at Marlow near Henley not far from my old school, so I went back to Marlow one day to take a look. The two bridges are indeed very similar in design, though the comparatively small Thames is no Danube, and the only similarity is brown murkiness. I had a good lunch in the *Compleat Angler* pub in Marlow in sight of the bridge, just to make sure of the design features.

I gave it a great push in the business sense during my early visits to Hungary, mainly because I loved to visit Budapest. During one visit a Hungarian friend gave me tickets to the newly refurbished Opera House, but alas, my travelling business companion was a true Philistine, so we left 'The Marriage of Figaro' during the Interval and headed off to a Bar instead. Perhaps it was the heady combination of Italian lyrics with digital Hungarian sub-titling which did for him. I parted company with him to city-walk and to get more of a flavour of the place and there were interesting antique shops and delightful architecture plus of course the usual Platoons of Hookers patrolling or waiting in ambush close to the Intercontinental. I'm not judgmental on the subject, people do what they have to do.

Business was slow though and nothing much crystallised during those early visits; I guess it was too early in the NATO/EU accession cycle. Some Defence Companies could not quite bring themselves to sign early deals in Eastern Europe and many Eastern European Users were still in an exploratory mode. Sometimes you must take the longer view and marshal your contacts for later activity, although one French Defence Company I freelanced for were quick to engage there where the Brits' would not. Someone was clearly doing big Defence business in Budapest though as the Arms Cash spectre raised its beguiling and familiar head yet again, sending the Hungarian Military Prosecutor's Office into overdrive.

Over a ten-year period, criminal activity in the Defence domain embraced eighteen individuals, four Hungarian Ministry Companies, and a whole raft of private Companies in a series of Hungarian Arms Cash events. A senior Official indicated that at least four Generals and around a dozen other Officers had been involved in Arms Cash activities delivering some 200m Forints in local kickbacks, according to the Hungarian Paper *Népszabadság*. Maybe that's why there was hesitation in getting even straight deals on the table, everyone had become suspicious and Institutional rigor mortis had set in.

I was followed back to UK several weeks later by an aspiring Hungarian Agent who wanted a UK aerospace Company to bid for a MiG refurbishment programme and characteristically, he would not take no for an answer. I sat with him and took him through all the reasons why the Company would not or could not undertake such a programme, especially since experienced local Competitors were much better placed than our Company. Maybe his winning formula was not based upon technical competence, maybe it was going to be just an Arms Cash proposition. Probably.

His timing was bad though, because we'd become extra wary about Agents of any sort by then and when he demanded over forty thousand Sterling just to work with us, and this for something we could not actually do and did not want to bid for, I had to say goodbye. It was hard work even getting him out of the Office as he was persistent to the very end. I wished him well, but he wasn't happy. I recalled my old friend Ramzi's maxim '...*never make enemies in this business...*' but sometimes you just can't help upsetting people.

Chapter 7
Engulfed

Postcard from Kuwait City

I've sold a lot of Defence equipment and services into Kuwait but even prior to the Iraqi invasion it hadn't been my favourite place to visit by a long chalk. The atmosphere was always tense with undercurrents generated by political disenfranchisement, but local calls for some form of democracy were invariably silenced quickly. I'd been due to fly into Kuwait from Bahrain the day before the Iraqi invasion, but I was recalled by my Company abruptly and immediately without any explanation. When I arrived back in the UK, I watched the invasion unfold on TV. It was a close call for me and I've no idea to this day how or why the Company triggered my recall. I asked around at the time, but no one could answer me. I kept in touch with my Kuwaiti friends in their London Embassy and with my 'other' Kuwaiti contacts who'd evaded capture by the Iraqis, and made their way overland into Saudi Arabia, from where they established contact through our pre-existing cut-outs.

I visited the Kuwaiti Embassy often during the crisis to offer whatever support I could. Ironically the building was just a few hundred metres away from the apartment where I'd once met the glamorous Fatima and her Iranian Arms-buying friends. It was desperately sad to see Kuwaiti Officers seated around a TV screen as their Country burned under Iraqi occupation, and during one such visit a senior Kuwaiti Officer drew me aside and asked me if I could help obtain certain Defence supplies which were needed urgently. Substantial elements of the Kuwaiti Army had escaped apparently from Kuwait City to Saudi Arabia, where they fetched up sparsely equipped but thirsty for revenge. They had manpower and a few Tanks but were desperately short of basic equipment. The Officer gave me a list of needs (yes, yet another list!) and I promised to do what I could to help.

A Tube ride away, my Plc Defence Company Employer was sceptical and unsupportive. How, the MD asked me sarcastically, was the Company going to get paid for the two million pounds worth of kit which the Kuwaitis wanted quickly and where would I source it all? Didn't I know that all Kuwaiti financial assets had been frozen? But I was optimistic that I could source all the items and regarding payment,

I could tell the MD only that a very senior figure in the Embassy had given me his word and that was good enough for me. The MD laughed of course but then he'd never even visited, let alone lived in the Middle East and had no concept of Arab culture. My perspective was different. I'd lived and worked in the region and not just with a series of avaricious Defence Companies. A Kuwaiti General had given me his word that they would pay, and I trusted him.

The real disappointment was that I even had to explain to the MD of a major UK Defence Company, that on the assumption that the Kuwaitis returned to their homeland, they would remember the people who helped them when they most needed help and swiftly forget the rest. He was not inclined to help them and seemed to find the very concept quite comic. He actually laughed at me when I explained all this to him. The Kuwaitis had been amongst our very best paying customers for many years (we also had non-paying Customers) and they were famously the toughest negotiators too. The costs and risks of helping them now were miniscule in the great scheme of things and to support them would be both morally and commercially essential, and yes, I did use the word 'moral' in a Defence Industry context! Sorry about that, it won't happen again.

The Company didn't share my view initially, since for them the situation was an inarguable exercise in risk-aversion. It was bitterly disappointing to have to explain the geo-economic facts of life to the MD and other Board Members as I thought it would have been blindingly obvious to them and I said so. The MD's own horizon seemed to terminate at Manchester and anything beyond that was alien to him. It was one of the worst cases I've encountered of a lack of worldliness in senior Executives, but sadly it was by no means the sole example. The MD had never worked for any other Employer, he was completely blinkered and just kept climbing the same corporate ladder. How on earth do they appoint people like this? I've seen a multitude of them. It took a painfully long time and several high-level Corporate and MoD meetings to gain commitment to action, but eventually the MD and the Marketing Director were persuaded. I recall the MD's exact encouraging words to me at the time.

"OK then try it, but on your head be it Ralph!" Now there's enlightened, supportive management in action. The reluctance to supply an invaded ally whose cash we already held for non-delivered contracts was in total contrast to today's torrent of support to Ukraine and Israel. Very, very perfidious Albion.

I set about assembling the deal but just about everyone else in the Company backed away from me as if I had Plague. The MD's attitude seemed extra-rich when I considered that we already held a substantial down-payment on an unrelated, and unshipped consignment for Kuwait which we could not fulfil owing to the Invasion. The Company had not even manufactured the products for which this down-payment had been made so we were significantly up on the deal already (similar principle to the huge Iranian down-payment for Tanks, which were neither manufactured nor delivered).

At one point the Company even sent a dysfunctional Scottish Commercial Manager to the Kuwaiti Embassy to remind the Staff there, who were watching on TV as their country burned, that they owed the Company money (a trivial sum compared to the advance payment the Company held)) for another contract item. It was toe-curlingly embarrassing and I had to run around hard and fast to patch up the damage this Presbyterian idiot caused. The General asked me to keep him away from them in future for his own safety.

Some of the equipment the Kuwaitis needed was easy to source, but other items would have to be bought back from UK MoD stocks, and this presented another challenge. The MoD was nervous about selling from their own stocks lest they needed these themselves for a War with Iraq which was yet to commence. The stock holdings included a veritable mountain of kit and munitions held in UK Depots, in Germany and elsewhere. Enough in fact to fight off the whole Soviet Union twice over and then some. What I needed for the Kuwaitis was a drop in the ocean relative to these massive over-stocks which had been accumulated by the British Army at Taxpayer's expense over many years. Today the opposite seems to apply, and we have apparently nearly run out of munitions without actually fighting anyone.

I attended a series of MoD and Government Agency meetings as a Supply plan was formalised to gather up all the required equipment as quickly as possible, then get it all aboard a UK Military Logistics vessel bound for Saudi Arabia on behalf of the Kuwaiti Forces. The schedule was very tight as UK Forces too had materiel to move around and it was hard-going administratively. I had to tread on lots of toes to get this all moving, so I did.

It was going well enough until MoD co-operation suddenly paused then ceased abruptly and even went into reverse. Having been persuaded to be supportive the MoD now became obstructive and resistant, almost a case of reverting to type. I persisted though and each time an MoD obstacle was presented, and there were many of them, I found a way around it. Finally, with minimal help from the MoD and not much more from the Company either, all the stuff was assembled and delivered to Marchwood Military Port for out-loading to the exiled Kuwaiti Forces.

It was a huge relief to get it all done and I happily hand-carried the Bills of Lading personally to the Kuwaiti Embassy in London, as payment was due on presentation of these documents. Embassy Staff were delighted, as was I, even though I was irritated by MoD's abrupt shift in co-operation from being very co-operative initially, to 'f*** you'. Now that all the kit was in Marchwood Military Port I was determined to find out what was really going on behind the scenes as invariably there usually is something going on. MoD would have to explain themselves. I dragged my MoD 'soft-ear' out of his Office to lunch in Topo Gigio's (sadly gone now) and poured red wine into him until he told me the inside story.

The Kuwaitis understandably expected Saudi Government support but at that stage in events, and with the US & UK yet to commit openly to Military intervention, the Saudis sat firmly on the fence. They told a visiting British idiot Junior Minister that they'd yet to decide which horse to back, Kuwait or Iraq, and didn't want to be rushed into deciding simply because some avenging Kuwaitis were hell-bent on fighting for their homeland.

The Minister was advised that the Saudis fully expected Britain to be asked to help the Kuwaitis, which would be natural enough, but for the time being the UK Minister was urged 'not to be too helpful' to the Exiles. After all, if the West wasn't going to throw out the Iraqis any time soon then Saudi Arabia would simply make friends with them. Once this exchange of views had been absorbed, diplomatically interpreted then disseminated back in the UK, it translated as 'don't help the Kuwaitis' and the impact was immediate. It produced the unexplained drop-off in assistance to people like me who were trying hard to help their Kuwaiti friends. Perfidious Albion was very hard at work once again. Notwithstanding all that, the ship finally sailed with all the equipment on board.

On the liberation of Kuwait City, I was flown into Kuwait via Saudi on an RAF C130, having firstly been rushed to RAF Brize Norton on a Company Charter plane. The Kuwaiti sky was black with the smoke from the burning Oil Wells fired by Saddam Hussein's retreating troops and I had a Brief to secure more Defence business. More Perfidious Albion on show here I'm afraid, grudgingly help a Nation in deep trouble, but only if it suits you, and when you get the 'all clear' ask them to buy your over-priced Defence products and services, whilst also charging them for clearing up the battle debris you dumped on them, including your Depleted Uranium dust.

Ah yes! but what about the outstanding payment for all that Defence materiel we finally managed to supply to the Kuwaitis in exile via Marchwood Military Port I hear you ask? I'd gone back into the Kuwaiti Embassy with all the Bills of Lading and documentation for the shipment and re-joined the Kuwaiti Officers there as they continued to watch the CNN reports from the Gulf. As I took coffee with them, I didn't have the heart to ask for the money, but I didn't need to. After coffee, the General drew me into his Office.

"Aren't you forgetting something Ralph?" he asked. The General opened his desk drawer, withdrew a cheque book and proceeded to write a cheque in payment for all the equipment. It was a big fat cheque and I watched him sign it.

"Don't worry..." he said "...it won't bounce, it's a very special Account." I didn't think for a moment that it would bounce, frozen assets or not, and I took it back to the Company's London HQ where they didn't know what to do with it. NDC's Corporate HQ had no means of handling cheques so the CFO and I walked it over to the nearest HSBC Branch where it raised a few eyebrows.

Success in this unusual situation had been severely hindered by the MoD's Defence Export Services Organisation and was certainly not enabled by it, but I bet they all boasted about it after liberation. No doubt the FO was quick to cite it as a tangible example of British support to Kuwait, but as to whether the deal was subject to Arms Cash elements I cannot say.

Kuwait was famously awash with Arms Cash deals until a reforming wave broke over the Parliament after liberation and one early outcome was an investigation into a $1.1bn Airbus Caracal military helicopter deal with France, in which it had been alleged that an Agent had demanded a commission of over $70m. I was happy to be well clear of this latest controversy as Kuwait never featured highly on a list of my favourite places to go even before the Invasion.

When I'd been flown from Al Khobar in Saudi Arabia into Kuwait on liberation of the City, I was flown low over the vast burning oilfields so I could see the nature of the next task to come. It was midday but the sky was black with acrid smoke, daylight could hardly penetrate those toxic clouds and deadly orange flames leapt high beneath them; it was truly a vision of hell on earth.

Saddam Hussein's departing Forces had torched the Wellheads and sowed Mines around them to prevent Firefighter access; a burning Oil-well is hazard enough without the added lethal threat of Mines. The first task would be clearing a Mine-free path to and around the Wellheads. I stayed in the City centre in the scorched but serviceable Holiday Inn and all the Spooks, opportunists and Carpetbaggers in the world descended on Kuwait.

A rare political Opposition leaflet appeared briefly in the City and I even saw copies of it in the Hotel. It demanded political reform and democracy for Kuwait, but this optimistic dream was swiftly extinguished, and the ruling status quo was reinforced and sustained. The much-vaunted high-value Western construction business opportunities to re-build Kuwait didn't materialise, as the City, though scorched like the Holiday Inn, remained largely intact and functioning in contrast with Iraq in the next conflict, which the US primarily and UK forces wantonly and unnecessarily smashed to a pulp. Sow as ye shall reap.

An Embassy Staffer took me to the scene of the US 'Turkey Shoot on Highway 80', 'Death Highway' as it was called, along which the retreating Iraqi Forces and civilians fled, desperately heading for home. A lengthy winding raggle-taggle mixed convoy of military and civilian vehicles carrying people and loot tried to crawl away from Kuwait. This was no disciplined military withdrawal but rather, a chaotic rout of demoralised conscripts with every man for himself. There was to be no escape for them though. The ramshackle, twisting column was caught in a succession of deadly US Air Strikes with no possibility of evasion and the result was mayhem and carnage. US Pilots, whooping over their radios, called it a 'Turkey Shoot' and televisual footage illustrated their description. Vehicles and people were smashed and burned in a horrific maelstrom with flanking minefields playing their deadly part in restricting escape.

The aftermath was an apocalyptic landscape strewn with mangled steel and bodies; the sickly stench of death loitered in the hot, still, air. The absolute mismatch between Coalition and Iraqi forces could be seen here in its starkest form, more so than at any other point in the brief conflict. The ageing and inferior ex Warsaw-Pact military materiel deployed by the Iraqis was in every respect outgunned, out-performed and over-matched by Western equipment and technology, whilst the complete absence of any opposing Air Force gave the Coalition carte blanche to destroy at will, which it promptly did with withering effect on Highway 80. Magnanimity amongst Victors had clearly become a thing of the past and this disgusting episode verged on being a War Crime. It should be compulsory viewing for anyone working in the Defence Industry, for this is what we contribute to and enable, and if a refresher was ever needed, then witness too, the subsequent wanton destruction of Gaza.

Postcard from Riyadh

I've visited Saudi Arabia on Defence business both as a Defence Company employee and on Freelance contractor assignments which included representing a Company with a range of Chemical and Biological Warfare (CBW) equipment. Although Western CBW equipment was developed primarily to combat the Soviet threat in Europe, Middle Eastern States were then practically and topically more aware of the deadly use of chemical weapons than most European States, as the lethal use of Chemical weapons in Iraq and Syria had demonstrated so disgustingly. A clear 'Customer need' existed for these products so I was pushing at an open door. Both Military and civilian demands for these products had grown at an unprecedented rate owing to anticipated Iraqi Chemical missile attacks during the Gulf War.

It was always said that in 'The Kingdom' some party or other usually had to be engaged for Arms Cash in Defence deals but this time I was completely detached from any Arms Cash factors, as I was working as a Freelancer purely to make Presentations. In this respect it was quite relaxing for a change, well, as far as anything can be relaxing for a lone Western Defence Contractor working in Saudi Arabia. On this visit I preferred not to know what was going on behind the scenes and it was much safer that way, especially in the home of the controversial and huge Al Yamamah Defence deal.

The infamous British Defence deal with Saudi Arabia has perhaps been the one of highest profile and highest value Arms Cash controversies in history and has been the subject of innumerable journalistic investigations and TV programmes. I'm not dwelling on it here as there is a tonne of reading and tele-visual material on the subject. From my perspective the most notable aspect was the abject failure of the Blair Government to pursue formal investigations into the deal when it had both a clear opportunity and clear moral choice as to whether to or not to do so. This failure highlighted and typified the fatal defects in the UK Government's position on Arms Cash generally and it was left to the USA to take a moral and legal lead, which was a huge irony. Here is an extract from a US Department of Justice Press Release dated March 1st, 2010, which sums it up well enough:

'In addition, according to court documents, BAES began serving as the prime contractor to the U.K. government in the mid-1980s, after the U.K. and the Kingdom of Saudi Arabia (KSA) entered into a formal understanding. According to court documents, the 'support services' that BAES provided according to the formal understanding resulted, in part, in BAES providing substantial benefits to a foreign public official of KSA, who was in a position of influence regarding sales of fighter jets, other defence materials and related support services. BAES admitted it undertook no adequate review or verification of benefits provided to the KSA official, including no adequate review or verification of more than $5 million in invoices submitted by a BAES employee from May 2001 to early 2002 to determine whether the listed expenses were in compliance with previous statements made by BAES to the U.S. government regarding its anti-corruption compliance procedures. In addition, in connection with these same defence deals, BAES agreed to transfer more than £10 million plus more than $9 million to a bank account in Switzerland controlled by an intermediary, being aware that there was a high probability that the intermediary would transfer part of these payments to the same KSA official.'

Whatever Politicians might say or claim, there is no UK Government moral position on Arms Cash or Arms exports and the sole advocate of an ethical British Foreign Policy, Robin Cook, died many years ago. Britain may now 'enjoy' some of the self-proclaimed most rigorous anti-Bribery legislation in the world, but it has no value without implementation and the will to implement. One or two minnows get targeted and caught from time to time 'pour encourager les autres' but major Company offenders have been allowed to swim free under the Government's nose. The UK Government is currently exporting weapons into two War Zones and helping to sustain conflict in both; where is the moral high ground in these cases?

In 2019 it was reported that the UK Serious Fraud Office had requested approval for charges to proceed against the Airbus SE subsidiary 'GPT Special Project Management', which was alleged to have paid bribes to win a Saudi National Guard contract and in which the MoD apparently had technical involvement and it was reported at the time that the UK Attorney General sought independent advice on the matter. Anti-Bribery campaign groups, including Transparency International, wrote to Sir Geoffrey Cox, the Attorney General, raising their concerns, but Airbus later announced that it was closing-down the subsidiary so any potential prosecution of the Company in question seemed to be disabled. But in November 2023, the Financial Times reported on opening statements made at Southwark Crown Court, relating to alleged corrupt payments made in respect of a contract with the Saudi Arabian National Guard (SANG).

The contract had been for the installation and operation of military communications networks and the Prosecution stated that 'Deep corruption' was at the heart of the case against two British men, one an ex MoD employee. Both were accused of bribery in Saudi Arabia according to the Prosecution. Reporting restrictions (i.e. censorship akin to Israeli gagging Orders) were applied by the UK Government and one of the defendants was also accused of receiving kickbacks. The history of the case stretched back to the 70's, but some of the alleged offences were said to have occurred between 2007 and 2012, so it took an awful long time to get this into Court, and when it finally got there reporting was banned. In December 2024 the 'Spotlight on Corruption' website reported:

'The men were acquitted of corruption in March 2024 after they argued that the payments were made with the full knowledge and authorisation of the Ministry of Defence (MOD).'

I smiled at the irony of the Trial location, Southwark, south London, once also the site of the MoD's huge St Christopher House offices where jailed former senior MoD Official Gordon Foxley had worked, and where the Iranian Tank contract, awash with Institutional Arms Cash, had been managed. It had been my workplace too as a young Contracts Executive. The Saudi GPT story has now been widely reported, so nothing new here, but the recurring theme was that no real action resulted at the time it was first exposed, despite a lot of very high-level prompting and lobbying. A casual observer might wonder why successive UK Governments experience such difficulty in dealing with Arms Cash situations where the common ingredients were constant and replicated. Back in 2006 it was reported in the Financial Times that:

'Interference in Middle-Eastern corruption cases is a sensitive topic for the SFO, which saw its investigation into BAE Systems' arms deal with Saudi Arabia narrowed after intervention in 2006 from Downing Street on National security grounds.'

This use of 'National Security' as a basis for stifling investigation is reminiscent of the Israeli technique and use of gagging mechanisms to suffocate domestic Defence corruption investigations. Even the famously secretive Saudis seemed to have made more progress than the UK on major bribery events. According to the SPA State News Agency a Saudi Defence Ministry Official was arrested in 2018 on charges of receiving a $260k+ bribe and the Kingdom's Attorney General, Sheikh Saud al-Mujib, was quoted as saying:

'The official sought to facilitate irregular procedures for the disbursement of financial dues to a company, taking advantage of his professional influence.'

In 2018 a group of elite Saudi business figures were arrested and held in the Ritz-Carlton in Riyadh during Crown Sheikh Mohammed bin Salman's apparent purge on corruption, although his action probably had additional political motives. In 2019 Spain joined the Saudi anti-corruption chase when a Spanish legal Investigator accused senior Officials of corruption and money-laundering connected to Spanish Arms sales to Saudi Arabia. The Spanish Court accused nine Executives of 'illegal conduct' over 11 contracts worth more than

$55m. The goods in question were Tank, Small Arms and Artillery ammunition, so good old-fashioned hardcore Arms Cash. Ah! I remember it well.

A subsequent visit to Riyadh to market Military Simulation and Training Systems for a British subsidiary of a US Company provided an altogether different experience of the Kingdom. This time I worked through an open local Sponsor Company which arranged all the appropriate Customer meetings and briefing sessions. After fact-finding meetings with senior Air Force Officers, I spent a week in my Sponsor's Offices with a colleague to write a Presentation and Proposal for a new Training Facility. We presented this to a senior Air Force audience, and we were invited to submit a formal Proposal. We subsequently won a significant order with no Arms Cash and no complications or threats. It can be done.

It used to be said that Saudis' massive Arms purchases were mainly made for Arms Cash reasons, but the deteriorating relationships on their Yemen Border changed all that and controversial attacks on Yemenis using British and other Weapon Systems have sparked outrage from several Human rights organisations and neighbouring Arab States alike.

A 2019 report by *Mwatana*, a Yemeni Human Rights Organisation which investigated 27 Saudi-led attacks in the 2015-2018 period, confirmed that laser-guided Paveway IV bombs manufactured by Raytheon in USA and Scotland were used on tribesmen with 122 women and 56 children amongst the victims, and overall the twenty-seven airstrikes caused 203 deaths and c. 750 injured. In a major political about-turn in January 2024 the German Government lifted its opposition to additional Eurofighter Typhoon jet fighter sales to Saudi Arabia by Britain, having previously blocked them owing to Saudi's questionable involvement in the Yemen Civil War and it's appalling human rights record. Now there's Arms Cash morality at work.

Today it's virtually impossible for any British-based Defence Company Sales Executive to respond with a Proposal to a Customer Defence Requirement whilst in-country as I did then. Companies have adopted disabling multi-layered internal 'Bid-processes' which are grid-locked by risk-aversion protocols which disable any agile response. Stilted commercial Compliance mantras suck out all creativity and imagination so now 'back-channel' arrangements are back in fashion to sustain momentum. It just takes too long to get any Proposal or even a simple Quotation on the Customer's table these days and Customers get fed up waiting and buy elsewhere.

I've worked for some of the world's leading Defence Companies but none of them have any form of front-line Sales or Commercial 'quick-close' capability so I love competing against them. In my experience, British-based Defence & Aerospace Companies have lost billions of dollars-worth of business and continue to do so, simply by being unable to respond when asked. My new, open Sponsors in 'The Kingdom' were very happy with our contract win and sometimes I met up with them off-stage when they drove across the King Fahd Causeway which links Saudi Arabia to Bahrain, 'Fantasy Island' for relaxing weekends, but Bahrain was a strange place for other reasons.

Postcard from Bahrain

In contrast to ever-tense Saudi Arabia and Kuwait, my visits to Bahrain were usually pleasant, except for the occasional life-threatening 'phone call of course and the ready availability of alcohol makes Bahrain regionally magnetic to many. Bahrain is a tiny State, barely three and a half times the size of Washington D.C. according to the CIA World Factbook, and it became free from Britain's 'Protection' as recently as 1971. Not that anyone would notice that now, because in 2023 the USA signed a strategic security and economic agreement to expand Defence and Intelligence collaboration with Bahrain. There were practical limits as to what a UK-based Defence Company could be achieve in Bahrain because of the modest scale of their Defence Forces, but now it's all changed anyway as the US is the top-dog and that's pretty well the recurrent theme all around the Gulf region.

Bahrain was always a great place to relax and Bahraini Defence Officials were ever welcoming and engaging, so even though I only ever managed small scale business there I was always happy to go back. The real tensions in Bahrain were internal security issues and especially the pro-Iranian elements which sought to oust the local Royal family and establish Iranian dominance. These fissures erupt openly from time to time and lethal incidents and attacks take place there, so take care when out and about.

Bahrain was deeply impacted by the regional geo-political tectonic shift to the USA after the Gulf Wars, when it agreed to basing rights for US Forces. It became home to the HQ of the US Fifth Fleet which is a giant-sized military presence in such a small city-State, so Bahrainis presumably feel very safe now. Bahraini Citizens enjoy a higher per capita GDP than those in the UK, but of course, there are only 1.5m of them. The old British Naval Base *HMS Juffair*, little more than a Quayside and Jetty really, was re-established in 2014 then upgraded in 2018, when that pillar of Naval strength, Prince Andrew, represented the Nation. Ah! what reassurance that must have given Bahrainis and the Royal Navy alike; Nelson would be proud. More Empire posturing. Transparency International once said of Bahrain *'Corruption risk is particularly pronounced in military operations and defence procurement, where anti-corruption safeguards are weak to non-existent'*…

…which sounds to most Arms Cash Practitioners more like an invitation than a criticism. I've never actually met any 'Transparency International' Staff in Middle Eastern locations, but I'm sure they're all experts. Maybe it was the ripples generated by the 2011 Arab Spring which triggered US concerns over the security of their Bahrain-based personnel which prompted Bahrain to spend, spend, spend, on even more US defence and security materiel. The tiny State blasted through a staggering $6bn worth of Defence purchases from the USA. Bahrain is lowly placed in the Gulf relative-wealth league table, so this high-spending level strained every financial sinew in the Country, which subsequently went out to neighbouring Arab States for loans. A mountain of those famed Bahraini Pearls could not buy all the jet Fighters the Bahrainis wanted.

My keenest personal memories of Bahrain relate to the chase for the missing NDC Arms Cash millions, since it was in Bahrain that my Corporate sheepdog, Harry Kensington, and I, met-up with Hector to try and resolve NDC's missing Arms Cash situation. On a later visit I was recalled abruptly from Bahrain by my Defence Company employers when I was about to fly from there into Kuwait, just before the Iraqi invasion occurred. The US military presence in Bahrain was already building up rapidly then and I was struck particularly by the sight of an Amazonian US Flight Mechanic atop some mobile steps wielding a giant-sized wrench, with which she seemed to be singlehandedly bolting a tail unit onto an F.16. But aside from that, my lasting impressions were of streams of Saudi cars pouring across the Causeway to 'Fantasy Island' carrying partygoers to drink away their weekends in Hotels, a practice which is never going to feature during any weekend within the Kingdom of Saudi Arabia itself, except in the idiotic illicit Stills of the Expat' Worker Compounds.

No getting away from it though, Bahrain remains a repressed Police State awash with Human Rights abuse. But when the Bahraini Crown Prince made a secret flying visit to Britain in 2023 with his Prime Minister, it was apparently to inject £1bn into what the Bahraini Press described as the *'exhausted British Market'* but cynics say this was blood money to assure British silence on Human Rights abuses. Rishi Sunak was advised at the time that *'The UK government must not allow itself to remain complicit in human rights violations in Bahrain'*. Whoops!

Britain sold Bahrain a modest reported c.£58m worth of weapons etc., between 2008 and 2017 which is small change in the great scheme of things, despite numerous loud, enduring and informed calls to suspend Arms sales to Bahrain. Never mind though, Prince William was in on the photo-shoot opportunity with the Bahraini Crown Prince, so all's well in Royal-tone-deaf-World.

Unrest breaks out periodically in Bahrain, just as it does predictably and inevitably under any form of repression, and such occurrences illustrate just how unstable some of the small Gulf States truly are; it wouldn't take much to pass the tipping point. In several, probably most of these small States, the autocratic status quo is maintained solely by an informal system of economic patronage in which Arms Cash undoubtedly plays its part.

From a personal perspective I found Bahraini GHQ Staff friendly and professional during my visits, but at that time, bigger richer States had priority on my Corporate agenda and Bahrain could be a lot of work for limited reward. It's cyclical of course, as today's top National buyers are tomorrow's Sales black holes. The US pretty-well owns the place now so no need to hurry back there, since if you're not American you won't sell much there by way of major systems and UK ambition has sunk to sweeping-up the product chaff. It was like a stopover on a bigger trip. Next stop neighbouring Qatar.

Postcard from Doha, Qatar

Doha was a quiet place when I very first visited in the late '80's. LPG exports had yet to come fully on-stream, urban development had yet to accelerate and the Gulf War had yet to occur. I would hardly recognise Doha today if left to wander around there, such has been the pace and extent of development. I plied my trade in and out of Doha in those early days, promoting a variety of Defence products, but times and methodologies have changed since then. If ever a modern Defence Company website statement struck a chord in a hardened Defence Salesman's heart (if you can find one) for what it doesn't say, then this is surely it:

'BAE Systems is working in partnership with local companies to deliver sophisticated defence solutions across multiple domains for the Qatar Armed Forces.' H'mm.

Qatar bought some token Eurofighters and French Rafale jets (there was ever an appetite for French military aircraft in Qatar) probably as a political consolation prize for Europe to counterbalance the 10,000+ US personnel, plus the U.S. Combined Air Operations Centre, plus the U.S. Air Forces Central Command, plus the U.S. Special Operations Central Forward Command and Central Command's Forward Headquarters, all based at in the same huge Al Udeid Base, to which the Qataris committed a staggering $1.8bn in expansion finance. This massive Base was the post-Gulf War lovechild of a Defence co-operation deal signed with the USA in 1991. It's a sort of different take on Arms Cash really, you pay for your protective tenant's Accommodation. It's akin to a National-scale Protection racket. But Al-Udeid is a bit too close to Iran for US comfort now, so after the Israeli attack on Iran the US hastily removed squadrons of its valuable Jets just in case, and sure enough, a few Iranian missiles arrived shortly afterwards. I bet that reassured the Qataris.

The Qataris, there are only three million of them, really need to keep the West, i.e. the USA, on-side these days, because they fell out with their neighbouring Arab States, some of which imposed a blockade because of Qatar's 'relationships' with Iran, which has now fired a few missiles into the Country. It's hard to keep everyone happy! Few, if any other countries, could match Qatar's one-time increase of 434% in the National Defence Budget and US Company Raytheon became a prominent Qatar Supplier, winning a $1.1bn contract for building an Early Warning Radar System in the Emirate. But in 2021, The Wall Street Journal reported that...

'U.S. authorities are investigating whether payments by Raytheon Technologies Corp. to a consultant for the Qatar Armed Forces may have been bribes intended for a member of the country's ruling royal family, according to people familiar with the matter.' Previously a lawsuit alleging c. $1.9m worth of 'payoffs' had been dismissed on 'jurisdictional grounds.

In the post-Gulf Wars, post-ISIS world, the US airbase within Al Udeid has become the biggest most permanent US Airbase in the whole Middle East and Defence Industry pundits joked about Qatar becoming the newest State of the USA. Although Qatar is about as big geographically as, say, Connecticut, that State has fewer US Defence Assets based in it than does Qatar. US aircraft located in Qatar stand ready either to deter or to further escalate regional conflicts, depending on your point of view, or to be relocated hastily as reported, when a threat looms. Qatar's massive National Defence spending supports this simply huge US presence.

With Qatar tucked firmly under American wings, the Brits' have little Sales traction there, well, virtually no traction other than the Eurofighters, having frittered away influence throughout the region by all too often adopting a patronising imperialistic stance which generated resentment in the emerging demographic. It seemed to me that once freed from British influence and 'protection', all Gulf Defence Customers demonstrated a robust appetite for alternative national Suppliers, fobbing-off the Brits' with the occasional order to keep them happy, whilst also keeping political alignment options open in case Uncle Sam's Voters call their Boys home again one day.

Today's US:Qatar deep alignment is a far cry from the situation which existed when I first visited. Then I stayed in the innovative pyramid-shaped Sheraton Resort hotel which had been built for a GCC Conference. It was stylish then but compared to today's hotel architectural excesses around the Gulf it has become insignificant. Qatar was essentially a 'dry state', a scenario which never bothers me, but a British Diplomat with a local expat' liquor licence used to bring cans of Heineken to my hotel room, I don't know why, and unthoughtfully he left his empties in my room. My first visit to Qatar was thus a clean-up mission in more ways than one, firstly, to dispose of the Junior Diplomat's empty beer cans, and secondly to unpick a bad situation of a British Defence Company's own making.

The Company in question had sold sub-standard ammunition of dubious (Chinese possibly) origin to the Qataris and tried to pass it off as British. But local test-firing quickly revealed the obvious flaws, with some rounds barely clearing the muzzles of the weapons according to the Customer, and of course the Customer is always right. The Qataris rightly refused to pay for this junk but the supplying British Company simply ignored their appeals and consigned the project to 'File 13' on the optimistic basis that the problem 'would pass…' (they actually used those very words) and I was hired to resolve the situation. In Company files I found several desperate requests for assistance sent from Qatar, and originating at a very senior level, all ignored of course. I flew out to resolve the matter and did so eventually, but understandably there was limited purchasing activity after that. When a Company lies so blatantly to a Customer and then ignores his pleas for help, it can hardly expect further orders.

Another regional State overcame the ammunition supply challenge by building their own Ammunition production Facility. This exception simply proved the rule though that the small regional States had limited Industrial infrastructure and their modest consumption rates did not justify local manufacture. Less so now, as with burgeoning demographic and related employment challenges, some Gulf States grasp every opportunity for indigenous employment and demand inward investment and workshare from Defence Companies, in the form of Offset and partnership deals; perhaps the greatest mechanisms for concealing Arms Cash yet invented.

After Iraq's Defence-Industrial complex was smashed-up by the 'Allies', regional leadership in traditional Defence engineering reverted to Egypt and Jordan, which had always led the regional field. Neither country has the financial muscle or digital tech' resource to approach Western levels, so they have dropped further and further behind the West in the digitised era. The UAE has the financial muscle to participate and could simply leapfrog the usual development stages by buying-in all the necessary expertise and infrastructure. The growth in UAE-based local Companies with Defence interests has been remarkable. However, Defence Industrial development also brings its own risks as such Facilities become targets.

The advent and widespread usage of Combat Drones of various shapes, sizes and capabilities, has changed the equation yet again, with commercially available digital technology now available to be combined with improvised or low-end manufactured Drones. Iran has emerged as a lead supplier (to Russia) in this field but any of the Gulf States could become major Players if they chose to. It's just a matter of money. Remember though what happened after the British set up Iran with Defence engineering infrastructure. Sow as ye shall reap.

No doubt Qatar is now an exciting place to be, but there was something intrinsically dull about Doha when I was there. Bad timing by me probably, or maybe I'd simply arrived too soon before the LPG revenue boom, a couple of game-changing wars, and a World Cup tournament. My best achievements in Doha were clearing up the Diplomat's empty beer cans and fixing the dodgy ammunition deal, so not a penny of Arms Cash in sight.

It was a different Qatari cash connection which caught the British Press's eye in 2015 when it was reported that Prince Charles had been given multiple bags of cash during his meetings with Sheikh Hamad bin Jassim bin Jaber al-Thani, the former Prime Minister of Qatar. There was no explanation for, or transparency over this massive cash gift, in which respect it was a bit like Arms Cash. Also, there was no investigative action or unexplained wealth enquiry either, so that's exactly like Arms Cash. All's well that ends well then. I've no happy memories of Qatar, where more recently the negotiating elements of Hamas have parked themselves, so Qatar has become a focal point for Gaza negotiations. By way of contrast, Jordan always makes me happy and I would return there at the drop of a hat. Keep the bags of Arms Cash, I'm off to Petra.

Postcards from Amman, Aqaba and Petra

Who wouldn't seize any and every opportunity to visit Jordan, if only to race south to Aqaba, scene of the Arab Revolt's great victory over Ottoman forces, and just a (longish) drive from Petra, John William Burgon's *'rose red city half as old as time'*. Hard to imagine that a visit to Petra could be coupled with Defence business but it was, I managed it. The first time I visited Petra the place was virtually empty, save for the few cave dwellers who still resided there, but when I returned many years later, the cave dwellers had been expelled and the historic landscape was awash with happy-snapping orange-skinned Cruise ship tourists.

I understand the economic necessity for all this of course but Jordanians were richer in so many other ways without it all. Jordan may be poor by Saudi standards but the 'King Abdullah Design Bureau' is a regional leader in Defence engineering and Jordan has a long heritage of Fighter Jet operation. The late and much-lamented King Hussein well understood the value of military jets and Jordan was the first Arab country to operate them; the King himself was taught to fly by the paternal founder of the Marshall Aerospace Group during the 'Golden Age' of aviation.

The first time I flew into Amman I was on military Aerospace business, but I was distracted because Amman has such wonderful Roman ruins, the Citadel for example, within which is the famed Temple of Hercules. On a sour note I was stalked around the city for three days by a weird wannabe Agent before I flew down to Aqaba, and it was a bit discomforting until I resolved the matter. The business proposition in the south related to Military Flying Training and it was a great concept.

I wouldn't usually brief local UK Military Attaches on anything like this at such an early stage, or indeed ever if I could help it, but I had to on this occasion. The Company I represented insisted that it really, really wanted 'new-new' business but it backed-out of the project at the vital moment leaving the Customer, various local Companies, the Air Force, a major national Bank and of course me, high and dry. I give up. It happens a lot but it's getting hard to be philosophical about such idiocy.

So many UK-based Defence and military Aerospace Companies are insulated from international business reality by big, fat, soft UK MoD contracts and will never succeed in export markets, where they really have to fight for business or innovate to succeed; these guys should drop any pretence at exporting as they demonstrably don't really want to export, or simply can't do it, however much they say they want to. I've experienced Corporate 'run-away syndrome' so many times with different Companies in numerous countries and the problem is never at the Customer-end of the spectrum, it's always at the UK Company-end.

The Jordanians were, as ever, welcoming and business-like and budgets could be found or created for the right schemes, with Banks and Finance available to support good projects. It's British Executive shrinking violets who were the barriers to progress. I doubt that T E Lawrence would recognise today's pasty and skittish Executives as being British at all. Did Britain really build an Empire? Did that actually happen? If you want to see how to get things done at pace today, then visit the UAE. The Gulf region provides great examples of just how States flourish once released from the dead hand of Albion's 'protection'.

Postcards from Dubai and Abu Dhabi
I first visited Dubai on weekend Leave from Oman whilst I was serving in the Sultan's Forces. My simple choice then had been either to spend my rare free weekends in Muscat, which was unpleasantly cliquey with military 'Incomers' from Desert bases being unwelcome in local Expat' society (lock-up your ladies syndrome) or instead, to head off across the border to 'Razzmatazz-Central', Dubai. Even in those far-off days it was a no-brainer. It was a long cross-desert drive to get to Dubai from our remote Omani Camp and a special road pass was required for the journey. The first stop was Buraimi, the Omani Oasis enclave which is accessed through a slice of UAE territory, hence the special Road Pass requirement.

Buraimi boasts a modern re-creation of a Beau Geste Fort which housed another Omani Regiment, so it was a good place for a stopover. Sometimes I hitched a lift on an Oman Air Force Skyvan which linked army bases to Buraimi Fort, but mostly I drove there along the desert road. It could be a hazardous drive though, and our lovely Indian Regimental Doctor lost his family on this road in a Camel strike. Buraimi Fort stands adjacent to the UAE town of Al Ain, which became the consolation prize for me and my sidekick if there wasn't time to reach Dubai.

When it was an 'Al Ain-only' weekend we rushed like kids firstly to Baskin' Robbins Ice Cream parlour and thence to the Hilton Hotel, around which most of Al Ain's social activity whirled. As Omani military cars couldn't be driven any deeper into the Emirates than Al Ain, we developed symbiotic arrangements with local expat' Nurses who booked Dubai hotel rooms at a big corporate discount, and in exchange we did all the driving and organised the fuel; a simple exchange deal, I can't say any more than that. The primal limitations of our remote and sparse Oman desert Base endowed Al Ain with a sophisticated sheen by way of comparison, but Dubai was on an altogether higher astral plane. After spending months in the desert, going into Dubai was like going into Space. Welcome to the Pleasure Dome.

In the early post-British Protectorate days Dubai became a fast-developing frontier town awash with pioneering adventurous Expats and Chancers of all stripes, who were helping to build and develop the City and its Economy and making pots of money in the process. They worked hard, they played hard, but when they lost, they lost really badly. Today they've all gone, replaced with a new breed of dull metro-sexuals and corporate Expats who rant on about idiotic pastimes such as Wadi-bashing, as if the desert was some sort of giant Theme Park. I suppose it is just that now for most of them. When they built an indoor Ski Slope in Dubai, I knew the world had finally gone completely mad. Dubai boasts some of the ugliest modern buildings in the world and the city attracts sun-seeking orange Package Tourists from Manchester so it's cultural decline is complete.

Before Dubai transformed into the Disneyworld architectural playground it is today, the most memorable landmark in the city was the 'Clocktower Roundabout', and everyone knew this feature and could navigate to any point in the City from it. I doubt that I could even find it now, despite it apparently being refurbished as a 'historic feature'. I met a whole series of endearingly unbalanced people in Dubai in those early days, but later I just met Arms Buyers. Unbalanced too, but not endearing.

My preferred Hotel in those heady early visiting days from Oman had been the Hyatt-Regency, which is where our road journey with the Nurses terminated. I had my first taste of Gulf Auto-bling there and it was the first, and to date the only time, I'd seen a fully chromed Range-Rover. Now, even Dubai Taxis are Supercars and Ferraris are ten a penny. The place is gross. The Hyatt featured what could only be described as a pulsating Disco, the '2001 Odyssey' which served as a fizzing sexy social hub, but there were smaller music venues in Dubai too, like the Baker Street Club with its live Bands and earthy swinging atmosphere, and others too, all gone now, subsumed by oceans of wealth, tower-block Hotels and Offices. By the time I was back in Dubai on Defence Business, the maverick Frontier atmosphere had all but gone, replaced by giant-sized Corporations and platoons of hungry Eastern European or Asian Hookers in the hotel bars.

During my early Defence business visits to Dubai I stayed at the Meridien Hotel, which was then called 'The International' and which was close to the airport which is why I used it. The International was less glitzy and less glamorous than the city centre's branded Pleasure Domes, including now that phallic symbol of Gulf wealth, the Burj Al Arab, and it had the advantage of low construction, a couple storeys only, so escapable and good for personal security.

The International hotel stood on a pleasant green plot with a few palm trees and a convenient circuit road which served well for jogging, and it had the best 'real' weight-training gym in the City, along with the second-best swimming pool. I remember seeing Danni Minogue lounging there whilst on tour. The icing on the cake was that the International also served the best Lobster bisque in the Gulf and the Montecristo cigars in the high-end restaurant were prepared before lighting and thigh-rolled in Cognac at the table, by a gorgeous sexy lady called Grace, who knew how to flirt with her diners without being cheap, which is an art form mastered by few. Sadly, the 'International' got bought and developed so all its great features disappeared, including lovely Grace and her Montecristo's.

I toured the UAE frequently with my bag of Defence goodies as the massive US military presence ramped-up during the run-in to the first Gulf War. Never had so many US Service personnel been ashore in Dubai and it was a shock for the Visitors and Locals alike. I visited 'Sheriffs', a popular pre-Gulf War watering-hole, only to be greeted by the sight of a scantily-clad young Eastern European lady clinging desperately and sluttishly to a gyrating, heaving mechanical Bucking Bull, cheered-on by the howling cream of the US Navy. Within a year or two, Eastern European Hookers had become regular features in the Immigration queues to get into UAE. New money, new tastes.

The military assets of Dubai and Abu Dhabi which existed previously as separate Commands, GHQ in Abu Dhabi and Central Military Command (CMC) in Dubai, had been fused together but the two Emirates still distrusted each other. The Federal Capital Abu Dhabi was always the more conservative of the two cities, but Defence Industry Execs' were ever more eager to visit CMC when it still existed in Dubai. Dubai not only procured more military equipment than it was ever going to need or use but was simply a more fun place to be. Many speculated, rightly or wrongly, that some Dubai Defence procurement was destined for Iran, so it was surprising that Western Governments did not look more closely at the levels of equipment purchases. I was interviewed once over a consignment of US electronic Mortar Fuses which had apparently found their way to Iran, but thankfully. they were way-off my radar.

Today's mega-bucks Gulf Defence deals are done in Abu Dhabi where Defence procurement has become much more structured than in the days of my early visits, when business was more 'informal'. To give an idea of scale, the 1998 purchase of eighty F.16s from the USA was reputedly worth $6-7 billion and secured a reported 30,000 US jobs, whilst the 2017 upgrade programme alone for the same aircraft was valued at around $1.6 bn. In 2023 UAE announced Defence deals worth a further $6.3bn at their IDEX Exhibition, and at time of writing the 'on-off' US F.35-Predator-Smart Munitions package comes in, if it matures, at a reputed $23bn. The strong growth in local Companies in the Defence sector has been dynamic and notable and is important in a country with such a young demographic, where despite huge National wealth, Jobs remain politically important and socially essential. But progress is a double-edged sword, for young people in Gulf States are now comparing their indigenous traditional autocratic leadership systems with Western democratic models; hence it's important for them not to be unemployed and disgruntled at the same time.

The British lost their way in the UAE Defence market ages ago, and at one stage when I was still visiting the region, UK even withdrew their 'First Secretary Defence Supply' from the Embassy as it was deemed that UAE markets had irrevocably left UK orbit. It seems astonishing that one of the world's leading Arms-buying countries was deemed to be beyond British Arms sales reach. That this was true at the time was for reasons of Britain's own making as we messed them about, we were arrogant, and our very expensive products just weren't good enough anymore. Plus, we were obsessed with Arms Cash and they were not. They didn't have to be, we did. On top of all that, Britain had lost its international political clout, and everyone knew it; the USA carried the big stick whilst UK clung onto a twig.

UAE is still a young State in historical terms, having been formed as recently as 1971 from the pre-existing Trucial States, with which

Britain had a distinctive and protective Political relationship. But once free from the claustrophobia of British administration the UAE grew rapidly, and demonstrated the huge differences in vision, concept and ambition which existed between it and the UK.

British Defence Companies continued to peddle Defence equipment and military aircraft in their same old way, assuming that their UAE Customers would still buy the overly expensive equipment they were being told to buy, and British Arms Companies splashed their Arms Cash around to secure orders, a practice which increasingly offended locals. Emiratis were fed-up with being patronised by the British and coupled with the availability of new alternative Arms Suppliers with more exciting products and Systems, UAE market-share swung away rapidly from the British. The 1991 Gulf War hammered a big nail into the colonial Arms Sales coffin as US military dominance transformed from being simply a TV image, into a highly impressive on-the-ground local presence.

The televisual treats served up by CNN showed off deadly US Smart Weapons smashing through micro-targets at hundreds of miles range and these trumped everything else. Almost immediately Regional Buyers wanted only these high-tech weapons. Conversely, the coverage also suggested to some of the smaller States that there was little point in buying Defence Systems at all because the USA would step in anyway and take care of things. Trying to sell high-cost British Systems and products became an uphill struggle unless it was into some specialist niche or lubricated with Arms Cash. In the history of Defence Sales, there can be few more extreme examples of such a traumatic collapse in market-share and the Brits lost out. Such was their arrogance.

During the build-up to the Gulf War, Defence Salesmen like me were booked out constantly on flights around the region, but most Customers didn't want very much as most of them had already been oversold. The Gulf States neither expected to, nor wanted to participate in a war, after all, that's what the USA was for. So, in many cases the motive for buying switched from Defence necessity to political adroitness, with spending in the USA intended to guarantee future protective interventions. UK had nothing comparable to offer in Defence Systems or in comparable military intervention capabilities. Yesterday's men. Prime Minister Macmillan once said that the British had to be the Greeks now the Americans were the Romans, but in the Gulf, we now looked more like the Etruscans.

In most territories I've visited, and not just in the Gulf region, UK Defence sales efforts are frequently outgunned (excuse the pun) by the size, scale technological and political mass of the USA. US business culture is more aligned to the current generation of Arab businessmen in a way that the British equivalent is not. It is more 'can do' and not inhibited by scale. The US National vision for Arms sales also differs in practical ways as US Military Staffs in US Embassies are more assertive, better trained, more active and more highly focused on commercial Defence opportunities than are their UK opposite numbers, who remain plodding and amateur in comparison. The relationships between British Defence Companies and the UK Government is very poor compared to the equivalent US model.

Counter-balancing the American-dominated Gulf Defence sales equation is a regional mistrust of US political motives and enduring concerns about the US-Israel relationship, especially under Trump's leadership. But all this is more than offset by the real-Politik acknowledgement that the USA carries the biggest stick on the block and is willing to use it, as Iran has discovered. Successive UK Governments missed the key geo-political shifts that occurred in the Gulf and other regions and the unwillingness of some major British-based Defence Companies to adhere to and to respect local sensitivities caused irreparable damage. In memorable cases these included requests by Gulf States for British Defence Companies to desist from making Arms Cash payments, which some Companies promptly ignored.

A senior UK Aerospace Marketeer boasted to me once that he had been dragged into a Gulf Crown Prince's Office and told that if his Company made any Arms Cash payments for an upcoming deal, his Passport would be seized, and he would never leave the Country. The Executive assured the Crown Prince that no Arms Cash payments would be made when the reality was that a substantial Arms Cash payment was already arranged and later paid.

By way of delicious irony, this Executive's Company had been conned into believing that a particular local businessman was closely connected to and had influence with, the local Ruling Family, and Company Executives were invited to spy from their high-rise Office to see evidence of this relationship as the local Ruler visited the Businessman's adjoining Tower Office.

Comically gullible Executives crouched low in their office and peered out of their windows at the appointed hour and did indeed see the Royal Family member visit the adjacent Office block. They were impressed by what they mistook for final confirmation of a relationship and later paid an Arms Cash commission without realising that the visiting dignitary was actually meeting someone else in the same building and had nothing at all to do with the contract they were awarded. The beneficiary of the Arms Cash payment, who had worked-over the Company for some months before the office visit incident, pocketed the Arms Cash and left town. Driven on by greed, desperation and optimism, some Defence and Aerospace companies can really be that unbelievably stupid, but of course sometimes the Executives themselves are benefiting also.

I used to drive between Dubai and Abu Dhabi along the linking desert road, and over the world's biggest speed humps, for which Sherpas and Oxygen were required. Lazy folks could opt to fly the short distance instead and grab a few extra Air Miles. I once hired a local driver for this journey so I could work on a Presentation in the back of the car, but partway towards Abu Dhabi the driver fell asleep at the wheel, causing the car to veer violently offline and crash into the upright concrete block central reservation.

Had I not sensed the veering and shaken the driver to wake him at the last moment it could have been much worse. The impact with the concrete blocks smashed off a front wheel, but luckily the raised blocks also 'captured' the vehicle like tramlines and kept it in a straight line until it came to a halt in soft sand. We'd been traveling at considerable speed, so this tramlining effect undoubtedly saved our lives. I expressed my discontent vigorously once we'd extracted ourselves from the smoking wreck. By way of contrast I was once given a lift to Abu Dhabi in a gorgeous Bentley Mulasanne Turbo belonging to a local businessman, and neither noticed the speed humps, nor hit anything.

My early Defence Sales experiences in Dubai and the other Emirates offered a clear insight into a range of Arms Cash methodologies which I saw in action subsequently in other countries. Foreign Defence Suppliers were allowed, mandated even, to have official local representation in Dubai at the time of my first visits, and a small number of well-established family Companies dominated the Sponsor market.

These were respected local Businesses with major franchises, typically including Automotive, Construction, Oil, and Freight etc., although their loyalty as Defence Agents could never be guaranteed as they naturally wanted the best deal they could get for themselves. It was not unusual to find your Company Sponsor also secretly representing your biggest competitor. I discovered that at least one of my major technical Proposals was sent secretly to a competing foreign Company in Europe either by my Sponsor or by a competing Agent, and there were plenty of interactive Arms Cash deals to be had for those who wanted to play.

One infamous British Freelancer, who played all over the field was Jim Fife, a dour Scot and former British Army specialist who'd also worked for a Belgian Arms Company once. During his Army service, Fife had worked in Oman and the UAE where he learned the value and power of Insider knowledge in the Arms business. Fife hankered after the Arms Cash rewards he knew to be available, so once out of uniform and working for the Belgian Arms Company he cultivated Insider relationships within the CMC Procurement office in Dubai (this turned out to be a Clerk) and he used this channel to gain access to highly sensitive Defence pricing and specification data.

Fife's simple modus operandi was to make pre-arranged calls to his tame Clerk via a Public telephone from Dubai airport, whereupon the Clerk would meet him and hand over data from Competitor's Proposals. He provided Fife with hard copies of both pricing and technical documents, which Fife sold on to the highest bidding Defence Company in exchange either for Arms Cash or for an Agency Agreement through which he then earned Arms Cash.

Fife had no scruples (in the Defence Industry?) and often covertly represented several bidders at once to ensure that he couldn't fail to win. It was a mini version of the Ambassador Extraordinary and Plenipotentiary at work. He paid off the grateful Sri Lankan Clerk with a pittance and scooped-up bucket loads of Arms Cash from willing major Defence Companies and poured it into his Channel Islands Bank Account.

Fife was expert at manipulating Executives and setting them against each other and he was quick to compromise some with kickback propositions to ensure their loyalty. He secretly tape-recorded conversations and played them back to the employing Companies of Executives he either disliked, or those who would not co-operate with him. He tried this little trick on me once, but I was ready for it and it backfired badly on him. Fife was toxic in every respect and single-handedly damaged some Customer Procurement processes to an extent that operational Military capabilities were compromised. This impact of Fife's Arms Cash manoeuvring mirrored the UK corruption case involving Gordon Foxley, who corruptly awarded MoD contracts for Mortar Fuses to German Company Junghans, insofar as these actions too could have compromised Army operational effectiveness.

Some UAE Officers who were 'in the know' about Arms Cash Agent activities were dismayed at the levels of payments made by overseas Defence Companies, both to their local Sponsors and to their anonymous Arms Cash Agents who lurked in the shadows. One Officer drew me aside once to comment that '...*these people are raping my country...*' referring to the local Company which openly sponsored my Plc Employer at the time. I cannot imagine how he might have reacted if he'd discovered the full range of Jim Fife's activities. Fife died prematurely. I've no idea if his demise was assisted or not, there would have been a queue I'm sure. When I terminated a Sponsor Agent's Agreement once in UAE, the CEO who met me toyed with a revolver as we discussed the termination. It's that sort of Industry.

Winning Defence Orders in the Middle East is tough enough, but some Companies made it tougher still by shooting (sorry, pun again) themselves in the foot. I've prowled around huge remote desert Warehouses searching for handfuls of lost Parts, I've conducted Desert Trials to prove that special Truck tyres function in sand and I've delivered packets of tiny steel Springs to the Head of a Technical Directorate, all to resolve Company legacy problems which held up substantial Contract final payments. Finding the handful of little Springs for example, unlocked half a million pounds sterling in withheld payments.

I was an Observer once at a Middle Eastern Tank Ammunition Trial at which a British Technical Adviser fell out with the Customer's

Ballistics expert. They argued vigorously with each other on the Firing Point of the Trials Range and as the sun rose higher and higher so did the temperature of their argument. The issue was that the Customer's Tanks couldn't hit any of the targets with the Trial ammunition and the Brit' expert and the local specialist locked horns over how to resolve the problem.

The senior Officer present got so fed up watching this performance that he decided to shoot the tank gun himself, but he too missed all the targets, so now loss of face was added to all the other issues. The outcome was that all the very expensive Trials ammunition was expended without a single hit on any target and the Tank Commander drove off in a huff in his Tank, leaving the two 'experts' in the Desert arguing furiously over what went wrong.

Sometimes I flew into the UAE solely to meet-up with contacts from other Countries, just as I'd met the Former Ambassador who later threatened to kill me. The place used to be a haven for Arms Buyers, especially during the Iran-Iraq War, but now many people go to Dubai simply for Shopping Events or to crisp themselves in the sun. It has regressed. In those days I took *Abra* rides alone along the creek with no tourists aboard and I swam there too before billion-dollar Tower hotels blotted out the sun, and flotillas of super-yachts chocked-up the waterways.

Dubai combined a dynamic commercial Frontier vibe with an undercurrent of traditional values when I'd first visited from Oman years before, but that's all gone now, giving way instead to a money obsessed eyesore of a City, upon which Abu Dhabi looks down disapprovingly like a sensible elder brother. Few people seem to realise that there are seven Emirates in the UAE, but Arms Cash aficionados know that the regional big bucks Deals, and the new Defence Industry Investors too reside in Abu Dhabi.

Postcard from Muscat

I've visited Muscat on business for several different Defence and Aerospace Companies and for one Oil Company too; everyone likes to go to Muscat. I felt awkward the first time though because I'd served in the Oman Army and I really didn't want to meet-up and haggle with old friends across a negotiating table. Ironically, whilst working for my first major Defence Company I didn't have to leave London at all to win business in Oman, owing to the convenient London presence of the Company's friendly Oman 'Facilitator' who I met-up with at his London Club to pre-configure Bids and proposals.

Everyone loved visiting Muscat even if the perception of Oman held by many UK Defence Companies was totally out of step with time and reality. One excruciatingly patronising Memo from a CEO of a major UK Defence Company declared that Oman was '...*virtually British, so there should be no problem in closing orders there.*' He was wrong of course but his comment reflected the widely held British Company views at the time.

Some major British Defence systems were sold to Oman when the Country really couldn't afford either the equipment or the through-life cost of operating and sustaining them, aircraft in particular, and speculation was ever rife as to the reasons behind such sales. Oman wasn't a wealthy State in the context of Arab Wealth, so some of the glamour Defence and aerospace purchases looked more like vanity projects or worse to an increasingly restless and sceptical young population. People both inside and outside the Arms Trade knew about the Arms Cash activities of an infamous deceased former Minister and big and overly expensive Defence projects simply fuelled speculation amongst a new generation of Omanis that the bad old Arms Cash days were back.

It was a different take on the 'bad old days' which confronted me when a resident British Company Director invited me to Dinner at his lush Expat' residence in the Muscat suburbs. I was at the dining table with the other Guests when to my horror, our Host leaned forward to lift a little silver bell from the table and he rang it aloud several times. A liveried Asian waiter entered on command to serve the Dinner, and I felt as if I'd been transported back through time. Many British Expats in Oman were unchangeably 'Raj' in orientation, which is why many went there in the first place, they hoped to sustain this sort of life. Times have changed now thank goodness.

Oman was also the scene of one of the longest running international Defence Sales prospects I've ever encountered. A British Fighting Vehicle Systems Company had been pursuing an equipment opportunity for seven years and had spent a tonne of money on Trials and Demonstrations to no avail. They were fatally unwilling to let go of this Prospect which had been written into successive Sales forecasts by successive Sales Directors and cost a fortune to pursue. The Sales Directors concerned would have been well advised to read Dr Dixon's excellent book 'On the Psychology of Military Incompetence' since the incompetent military leader's characteristics identified therein, apply equally to Executives in Business. One of the key failings of 'the incompetent' is an inability to accept information or Intelligence which conflicts with their pre-formed view.

In this case, the pre-formed view was that the Oman Army would buy this wretched equipment, when in fact they would not and could not buy it for good reasons. A local Army friend of mine advised discretely what the problem really was. No problem with the equipment he responded, but the vehicles themselves were undriveable owing to a botched engine upgrade which left the driver's compartment too hot to use and the dispute with the Upgrade Supplier was going nowhere. Mystery solved. The Company Sales forecast developed a sudden gap, as had the Sultanate's borders.

Porosity of borders became an acute problem in all Gulf States, but especially in Oman, owing to lengthy, remote and inhospitable sections of Border. Drones are heaven-sent for patrolling such regions but it's still hard to sustain 24/7 coverage. During my Oman Army Service, it was a deployment to the northern Border with Saudi Arabia which was simultaneously comic and deadly serious.

The serious element was political, insofar as the Saudis had apparently demanded a land corridor across Oman for an Oil pipeline to avoid taking passage through the risky Straits of Hormuz. But tempers flared when agreement couldn't be reached between the Kingdom and the Sultanate. When the Saudis threatened to deploy troops, unheard of previously, I got the task of establishing Observation Posts (OPs) in the hostile Rub al Khali area to give early warning of any such moves by the Saudis, and our Reconnaissance Platoon was deployed by Hueys to this oven-hot and ill-defined Border area. This was one of those occasions when a stray shot or some accidental action can start a whole Conflict, so my fingers remained tightly crossed. With the OPs established I flew back to Battalion HQ for further briefings.

Twenty-four hours later a message was received in our COMCEN stating simply 'OPs *washed out*'. The encoded message was transmitted

219

in Morse and I asked for verification. Yes, came the affirmation 'OPs washed out'. Dawn was breaking as I took-off with a Patrol to investigate. Did 'washed out' mean 'wiped out'? Who knew? We dropped short of the OPs and approached on foot. It was horribly still and quiet which fed my imagination. We stumbled up dunes to reach the centre of the position where a real surprise was waiting.

The message was accurate. The OPs had indeed been 'washed out' not wiped-out. A freak seasonal rainstorm in the middle of the Empty Quarter inundated the area and caused flash floods. There were pools of water in the sand with sleeping bags and military bric-a-brac bobbing about in them. Morale had slumped. Disconsolate soldiers baked by the sun the previous day were sitting around in little huddles shivering, their expensive military hardware, futuristic Steyer assault rifles and advanced radios lay discarded, coated in sludge-like wet sand. Regardless of how well troops are equipped and or how ever well you plan operations, something simple like this can come along out of the blue and f*** you up. Every time I make a high-level Defence Product presentation I mentally recall this scene, just to calibrate my sense of perspective. Always remember to consider both the enemy and the rain.

Chapter 8
Any other business

Postcard from Seoul *via numerous UK stopovers and Golf in Somerset*
Sustaining jobs is oft cited by UK Governments as a justification for
Arms Exporting, especially in current times when Arms appear to be
the only thing the UK can manufacture, but it's sometimes Defence
Technology rather than the Weapons themselves which is in demand.
I once managed a transfer-technology package to Pakistan which
produced a rich and steady downstream business in components, and
it fell to me to escort a party of South Korean 'Senior-Industrial
Visitors' around a series of UK factories and Facilities with a transfer
technology deal as the prize. This Industrial tour took-in locations
from Scotland to the West Country and my Brief was to stay close and
to help them in any way possible.

I met the Visitors right off the plane from Seoul and set off with
them on a ten-day tour conducted from a hired luxury Coach with
integral Office facilities, not unlike a politician's campaign bus. A series
of UK internal flights had also been booked. Heading-up the Korean
Group was a diminutive but distinctive 'Boss' figure, who had to have
the best of everything and he was supported by a team of three
'Minions' who did exactly what they were told; certainly no more and
emphatically no less. In each hotel the Boss had a Suite, but the
Minions shared a room. The Boss had the pick of the seats and the
pick of the menus whilst the Minions made do as best they could.
Usually they took sandwiches to their shared room.

At an early stage in this Defence-Industrial magical mystery tour,
our happy band was joined by the Korean's UK Agent, Dave Strudley,
whose employment by them was both overt and permissible, so no
Arms Cash issues to manage. Strudley was a retired semi-professional
Footballer-turned-Businessman and was a larger-than-life character,
with a penchant for not merely loud but positively deafening Suits; Pa
Larkin meets Del Boy. His vocabulary was every bit as colourful as his
wardrobe. Technicolour.

What a prize-crew we were as we ventured forth with our 'glam'
professional Tour Courier, Lottie, for company. We'd arrive each day
at our programmed destination and be met by a Senior Executive.
There would be Boardroom briefings on the machinery or processes
we were going to see, then a Facility tour, lunch and a 'wash-up' session

before moving off to our Hotel for the night. The Boss was always led the Party with the Minions following at a discreet distance (Military term: 'a tactical bound'). They captured every spoken word during these visits and wrote them all down, regardless of whether Briefing Notes were issued by the Hosts. There were no classified items in the tour and the Minions scrutinised every machine and recorded all the Manufacturer's names; they paused at every sub-assembly and sketched things, they pulled and poked at components and wrote down parts descriptions, serial numbers and any other details they could find. Every visible and audible detail was recorded in the Minion's notebooks, they never stopped scribbling and sketching.

On day five we were due to stop overnight in the Midlands and I anticipated another dreary night in a 'Rep-Stay Inn' or similar, with Room notes about saving the planet by re-using the towels and a never-to-be-used Trouser-Press in the corner of the room; the Dubai Jumeirah Beach Hotel seemed a long way away now. But in fact, I was wrong about the next hotel. It was a lovely converted Country House standing in delightful grounds. This charming Period house had lost its historic Owners years earlier and had succumbed to commercialisation, probably as a result of UK Inheritance Tax or similar; so now it worked for its living, the architectural equivalent of a 'fallen woman'.

As we disembarked from our luxury Coach, I noticed the developing chemistry 'twixt ex-Footballer Dave and lovely Lottie, our forty-something and glamorous-in-an-Essex-girl-sort-of-a-way Travel Courier. Up to this point Dave had stuck to the Korean Boss-figure like pooh to a blanket, but now he was clearly distracted. It was the Boss's habit to enjoy some 'Me time' in his room (Porn Channel) at each stop, before descending to be entertained at the bar by Dave and me prior to Dinner. The minions appeared neither for Dinner nor for drinks, feasting instead on the sandwiches and soft drinks which were despatched to their room. They were always down early in the morning though and invariably ahead of the Boss, to gorge themselves on buffet breakfasts by way of calorific compensation.

That evening I found myself alone entertaining the Boss because Dave hadn't appeared in the bar. Dave's bottomless fund of tedious Soccer anecdotes always drew vigorous laughter from the Boss, even though the Boss clearly didn't understand a word of it through Dave's thick Brummie accent. Dave had a certain earthy style which much amused the Boss, in fact he could have been talking any old bollocks and often did, it made little difference. But if this made the Boss happy, and clearly it did, then that was fine by me. But in Dave's absence I had to improvise, and it was hard work keeping the evening moving along. In a fleeting moment of curiosity, I wondered where Travel Courier Lottie had got to.

I struggled on with painful conversational slowness entertaining the Boss, well past midnight in fact, and I was relieved to finally wish him goodnight as he headed off to bed. The Bar and Restaurant Staff had derived considerable entertainment from my laboured efforts to entertain him and they applauded after the Boss left the Bar. As I walked along the upstairs corridor heading for my room, I passed Dave's Room, and had I not already known this to be his Room, I would soon have learned as much.

That unmistakable 'Brummie' accent, or rather, those unmistakable Brummie grunts of exertion which emanated loudly from within made it obvious. Those manly baritone sounds were followed by an assortment of Essex soprano shrieks, groans and exhortations, such as '...harder Dave, harder...!' in the equally unmistakable tones of Travel Courier Lottie. The room's Period bedstead was clearly undergoing a relentless Consumer test and judging from the high-frequency squeaks which pierced the night, mechanical failure was a real and imminent possibility.

The following morning the Minions appeared for breakfast later than usual, looking for all the world as if they'd not slept a wink. I wondered if they too had been road-tested by Travel Courier Lottie, maybe she negotiated party rates. I just had to know so I engaged them in our usual halting banter, trying hard for the sake of international Diplomacy not to ask simply '...did Lottie f*** all of you?' It transpired that they'd not had this pleasure, but their room was directly beneath Dave's, who displaying commendable stamina, had sustained his efforts throughout the night, during which time the Period bedstead did indeed collapse.

Dave was obliged to dismount to effect running repairs and the ongoing noise, confusion and general excitement kept the Minions awake. Tired though they were, the Minions greeted Dave when he finally appeared at the breakfast table by standing and vigorously shaking his hand by way of congratulation. Dave was clearly wrecked and past caring, whilst Lottie missed breakfast completely and waited sheepishly for us aboard the bus, looking very much the worse for wear.

What about the deal though? Where was the business in all this? Nowhere so far. In the South West we checked into the final hotel of the tour, with an afternoon to spare after viewing a nearby factory. My own CEO, having carefully evaded all the actual work, was due to join us there. Stuck for ideas as to what to do with our Party during a whole spare afternoon, I struck up a conversation with the Hotel's Owner. This saint-like figure offered to get us onto the nearby Golf course where he was a Member; it was a faultlessly brilliant suggestion.

It was all plain sailing from there. I briefed my CEO and paired him on the Course with the 'Boss' whilst I marshalled the Minions who were not allowed to actually play, but who followed the Boss adoringly, as his 'Gallery', around all eighteen holes and applauded his every hook and slice, of which there were many. We all clapped and smiled a lot. By the end of the Round and after a really good dinner we had a deal; Golf had done the trick and a Technology package was on the table. It was with absolute relief that I packed them all off on a train to London for their weekend-off, before I made a run for it just in case, they changed their minds.

My heart sank the following Monday morning when the 'Boss' called me after postponing their scheduled return flight. He wanted to invite me to Dinner in London as a thank-you for my efforts during the Tour, and for positioning the deal with them. I prayed fervently that Travel Courier Lottie would not be on the menu too. I met up with them all once again in a Korean Restaurant in London and consumed vast quantities of Kimchi and Rice Wine and received an armful of well-meant South Korean Corporate gifts for my trouble. Dave joined us later but was unable to drink as he was apparently taking antibiotics. There was not a penny of Arms Cash in sight, although South Korea was far from immune from the Arms Cash virus as *The Korea Times* reported.

Several hard-core, high-level cases of Defence corruption hit the Seoul headlines at once, with two Naval Officers indicted of receiving bribes in one instance and as many as eleven other Officers arrested on other corruption counts. It was reported that a Marine Colonel had benefited by over forty million Korean-Won in one construction kickback incident, whilst in other cases, a Naval Officer intervened to help a fellow accused Officer, but only for a kickback of millions of Korean-Won for himself.

It was over a Reconstruction project in distant Iraq that it was reported that three other South Korean Officers were convicted of bribery and other offences in a programme financed with some $70m of American cash, and needless to say this event strained relations between the US and their South Korean coalition Allies, although the US had its hands full anyway with numerous corruption cases amongst its own military personnel in Iraq.

At a higher level in South Korea, the South Korean Head of Defence Procurement, Lee Myung-bak, resigned from his post, following allegations of corruption which was another clear example of the 'Influencer' at work. Across the China Sea it was another Influencer, ex-President Lee Teng-Hui of Taiwan, who was indicted on charges of embezzlement, involving sums of over seven million dollars-worth of the National Defence & Security budgets. This was of course in addition to that other regional Arms Cash classic, the *Lafayette Affair*, that long-running Taiwan Frigate corruption scandal which produced demands for Fines of over $900m against Thomson-CSF, now Thales, the French suppliers of the six naval vessels in question. These ships were purchased despite being around fifty per-cent more expensive than a competing comparable bid; a huge difference to be ignored in any selection process. According to Taiwanese Defence Minister Kao-Chun, additional Arms Cash kickbacks were identified in the subsequent supply of spares to support the controversial vessels despite an earlier prohibition order. Some things are just too hard to resist.

Postcards from Jakarta and Singapore *via Surrey, Seattle and Fort Lauderdale.*
We all make mistakes of course but this was a big one. My mistake was to join a small UK-based Defence Training and Simulation Company, Trio Defence, which had been bought-out by a Seattle-based Company called Alsco which sold Weapon Training Simulators in the USA. Alsco wanted the access to international markets which Trio Defence

afforded. It was all very logical on paper, a mini version of what had happened to much bigger UK Defence Companies and post-acquisition, Trio continued to be run by its two Founders as joint MDs (how can you have joint MDs?) reporting to their new Seattle CEO.

But Trio Defence had become vulnerable because these two MDs had overstretched the Company financially, leaving it desperately short of cash, and their operational management of the Business was weak to the point of non-existence. Trio had become wholly dependent on US Parent-company cash infusions for survival. I'd foolishly accepted the joint MDs' word that the Company was in good shape financially when I joined it, but as the saying goes, 'you can put lipstick on a pig, but it's still a pig,' and Trio Defence and Alsco were both covered in lipstick.

The twin Directors bought into a fantasy proposition from their US buyer, optimistically exchanging their whole Business and genuinely valuable IPR for an avalanche of worthless Share options. To compound Trio's operating difficulties, the two MDs clashed and were at each other's throats constantly; they were a toxic combination, fatally locked in an Executive Sado-masochistic relationship. Just days after I joined the Company a sombre-looking group of grey-suited men visited the Trio's Offices, and when 'the Suits' left I cornered Trio's young Accountant and interrogated him.

Nervously and reluctantly, the young Accountant advised me that the Business was in deep financial trouble and owed money to just about everyone. The 'Suits' had called in to arrange collection of what was owed, Tax and plenty of it. Only the monthly transatlantic subsidy from the US parent Company kept Trio Defence afloat. The Company was technically insolvent.

Trio Defence was well known in its niche Sector so there had been no outwardly obvious reason for concern; the twin MDs had assured me the Business was ripe and ready to grow. They lied of course. I should have known better. In the Defence Industry especially, neither good faith nor hope are bankable commodities, and once I knew the truth, I confronted the MDs about their deception. They were predictably and perfectly evasive, barely acknowledging the financial criticality of the situation, and what I most wanted to do was to punch their lights out.

The 'Technical' MD added value to the Business because he designed the bespoke Facilities sold by the Company, but he'd suffered a terminal double charisma bypass and was rude to everybody, including Customers, so he couldn't be allowed out of the Office without an adult. The other MD, the Sales and Marketing half of the pair, was a pompous, secretive little slug. His unwillingness to share information about the Company's international activities was a major issue and I distrusted him. He made it easy to dislike him.

How this dysfunctional pair had ever managed to set-up a Company together was a mystery, but they'd come up with a good concept at the right time, even if they lacked the business skills to exploit it. They'd wasted the fruits of early success with premature and optimistic expansion, so the operating costs of the Company had spiralled upwards unsupported by new Sales revenue. Their mutually exclusive and petulant egos clashed at every turn. Quite why US Alsco wanted to buy a Company wreck like this was a mystery too; surely US businessmen were too smart to buy into such a mess? Or was there a deeper, darker, smarter reason?

Contrary to my mental image of US business efficiency, the new Seattle Parent Company also turned out to be a mess. The Company's CFO, who'd been in the UK when I joined Trio, returned to the USA and promptly disappeared. I mean he really went missing. Behind Alsco's financial façade were two stereotypical US Financiers, 'Angels,' who were funding the cash-burning Alsco with 'OPM,' i.e. 'Other People's Money' from willing and clearly very rich American private Investors, who alone could afford such risky Investments. In Seattle, I quizzed Alsco's stand-in CFO about them.

"What are our Investors like Jenny?" I asked, wondering why anyone would keep pouring money into this corporate wreck.

"Oh, they're just regular American Investors, Ralph…. you know, rich, fat guys…"

I thought this was parody until I finally met them in Florida. They were Rich fat guys exactly, except one, a rich skinny guy. Well, it's the exception which proves the rule isn't it? So exactly why had Alsco bought this British basket case? I hoped to find out during my first visit to Seattle. Alsco paid the airfare because Trio had no money and it was the mother of uncomfortable flights, all bunched and crunched up with package-deal tourists in the wrong end of a heaving 747. Oh, so far away from my serene First-Class flight to New Delhi years earlier. Still, it was better than Zimbabwe Airlines, but only just. It was a close thing.

There was no 'meet or greet' on arrival so I found my hotel and prepared for my first meeting with Alsco's CEO. I pulled no punches with him, there was no point, Trio Defence was in deep shit which would only get deeper whilst the joint MDs were tearing at each other's throats and managing the Business ineptly. The Company's non-existent Order pipeline was an easy-looking challenge compared to the sectarian warfare which raged in the Board Room. I laid out the situation for the Seattle CEO, not warts and all, it was all warts. He listened casually, distracted by his constant checking of the DOW Index on the screen in front of him. Did Alsco do any Due Diligence? I asked. No response. I explained exactly how deep the shit was, then left him to think about it as I went shooting on Alsco's Company Range, unsure as to whether or not the distracted CEO heard anything I said.

Alsco's Shooting Range confirmed, as if any confirmation was needed, just how gun-oriented our American friends were, compared us Brits. A glamorous female shooting Instructor joyously pumped lead into a series of 'Hoods' in video Law Enforcement scenarios to demonstrate the System's capabilities. I think she climaxed with her second magazine. Scary woman. I felt the need to arm myself.

A redeeming feature of the whole visit would, I hoped, be a trip around Seattle which I'd yet to see, so I snatched a few hours and went firstly to the waterfront to visit Pike Place Market, where I watched brawny Fishermen heaving around huge Cod before a public audience. H'mm, was that it in Seattle then? Cod? There were a couple of interesting bookshops and other quirky small shops but not much more, so I wandered back into the City centre to take a ride on the Monorail out to the Space Needle. Seattle was a big disappointment as a City, sterile and superficial, apart from the old waterfront and market areas. Maybe the sour nature of the Alsco business had coloured my judgement.

I returned to the UK on a less congested though still cramped plane, not knowing if my briefing to the US CEO had any effect. By the time I landed in the UK I'd been appointed CEO of Trio Defence by email (which I hadn't seen). We hadn't discussed *that* in Seattle. The twin MDs read their copies of this email before I'd even seen it, so once I landed, they came a-hunting for me, all anger and belligerence and frothing indignantly. Then they stopped haranguing me momentarily and started on each other and had to be separated. Once back in the Office their desks were relocated to different buildings to keep them apart. The Company could manage very well without the Marketing 'Guru' MD, but not without the Technical specialist MD, but sadly their Alsco Shareholder agreements didn't allow for either of them to be fired.

I took urgent administrative actions to manage Creditors and HMRC and met with them to plan a solution. I also needed a new Accountant, for our youngster left for a more normal career and I could hardly blame him. I hired-in an Agency Accountant to get the Company Books together and executed a Business Review. We needed sales and revenue urgently, so I searched for a relatively quick win (anyone ever found one of those? No) and as I rummaged around in the Prospect database I ran across trouble in the form of an Indonesian Prospect.

Hard-copy details of this Prospect were held under lock and key by the Marketing MD and I had to prise the documents from his filing cabinet because he didn't want anyone to see them. I felt apprehensive opening the file as anything to do with Indonesia and the UK Defence Industry was trouble. Worse, US-manufactured Small Arms had been used in the suppression of the population there, including at the infamous Dili massacre, and there were widespread allegations of Arms Cash corruption in many Indonesian deals. In 2017, a '*World Peace Foundation*' article relating to Indonesia stated…

'Brokers and middlemen, frequently the conduit for corrupt payments from suppliers to decision-makers in international arms deals, are active in most procurement processes from a very early stage, creating ample opportunities for decisions to be distorted by corruption, from the setting of initial requirements to the final contract award. Despite this, before 2015 there were virtually no cases of military officers or defence officials being investigated in relation to corruption, and none in relation to arms procurement.'

British Armoured Vehicle manufacturer Alvis had been caught in an Agency firestorm there, when it was alleged that bribes had been paid by them to ex-President Suharto's daughter Siti Hardiyanti 'Tutut' Rukmana to secure contracts. A commission figure of £16.5 million was mentioned. A newspaper article stated...

'...allegations arose because a former Alvis agent, Singapore businessman Chan U Seek, sued Alvis over the Scorpion (AFV) sale claiming he was entitled to commission.'

Witness statements indicated that payments had been made to Suharto's daughter, but shortly afterwards, Alvis reached a confidential settlement and both sides stated they were *'...prevented from discussing the case.'* Another Guardian article stated...

The 100 Scorpion light tanks were sold with the promise from the Indonesian regime that they would not be used for internal repression. However, they were subsequently discovered in action in the breakaway province of East Timor and in Aceh. The sales were backed by the British government's Export Credits Guarantee Department, which was left to pick up a £93m bill when Indonesia ran into a financial crisis. President Suharto was ousted, and Indonesia has asked to postpone payment of its debts.'

Susan Hawley of 'Corner House' the anti-corruption Campaign, said the Government Export Credits Guarantee Department
'...should have spotted that the president's daughter was involved. Why didn't alarm bells ring?'
In true Defence Industry style, Alvis reached a settlement rather than fight in open Court with witnesses and evidence and in 2004 Alvis was bought by BAE Systems, so everything was alright again. This Arms Cash case was not unique in Indonesia, as there have been too many to detail here, but in addition to internal Arms Cash controversies, Indonesia had been subject to Arms embargoes and these impacted on what Trio Defence wanted to do in the country.

The EU Arms embargo on Indonesia was lifted in 2000 but a full US Arms embargo remained in force until 2005 when it was lifted by President Bush. Although no formal UK or EU Arms Embargo remained in place, the UK Foreign Secretary felt strongly enough about the situation to take a principled stand (surely unique for a Politician, probably a mistake!) against Arms exports to the Country, and Indonesia became a trading minefield for UK Defence Exporters, especially if lethal products or military training systems were involved.

Gaining formal Export approval in the UK was thus difficult if not impossible, but in any case, Trio's Indonesian prospect fell within the US Arms Embargo period and as a US-owned Company, Trio Defence was legally required to respect the terms of this Embargo. To complicate matters further, Trio Defence already had Indonesian baggage, because the Company had previously been targeted and caught out years previously by a TV Investigative team, which exposed their activities in-country. The issue of Trio Defence's Export Licensing on that occasion had been raised by a Parliamentary Select Committee.

This heady cocktail of US-ownership, an obligation to comply with two sets of National compliance regulations, a US Arms Embargo, an investigative TV programme, various in-country Arms Cash proceedings and UK political sensitivity, all combined to create the perfect Defence Company Arms Cash Clusterf***-in-waiting. All that was missing was an unscrupulous Facilitator. Hold that thought.

Given all this background stuff, most Companies would have cut and run, but Trio Defence suffered from those fatal twin weaknesses of desperation and secrecy. Desperation was easy to understand as the Company needed money, and secrecy? Well, that's a 'given' in the Defence Industry isn't it, it's always there, you just must work out why it applies in each instance. The Marketing MD kept the Indonesian project close to his chest and I guessed that Arms Cash was a motivating factor. Had a kickback been budgeted? Probably.

The MDs argued that Trio Defence's bespoke Training Facilities were in themselves benign and didn't kill or hurt people, so the embargoes didn't apply to them. In the literal sense this could be true, but it was pedantic. Trio Facilities were inert constructions yes, just as a rifle is an inert fabrication, but the capabilities derived from them facilitated killing, and evidenced by contemporary events, Indonesian Security Forces did not shy from using lethal force.

I doubted that the Company would get an Export Licence for this Prospect so for a change it was a 'no-brainer' for me to go to the UK Defence Export Services Organisation (DESO/DSO etc.) to ask that very question. My determination to check the export situation formally and officially was neither shared nor welcomed by the Marketing MD, who didn't want it discussed outside the Office. The whole thing stank to high heaven. Trio's previous Indonesian business had attracted TV Journalists and Parliamentary attention, but it wasn't either this or the Human Rights issues which really caught my eye next, for Trio Defence had an Agreement for Indonesia with a shady British intermediary Company called 'Whites', which was already doing low-level security business there.

I obtained a copy of the inter-Company Agreement between Trio and Whites and the draft Supply contract which it enabled. The Agreement and Contract documents, and therefore the plan too, were convoluted even by Defence Industry standards, and despite my own MoD Contracts Officer background and extensive Defence export experience, it was a challenge to unpick it all. I sent it off to a respected Commercial Lawyer for independent legal analysis and advice.

In brief, the arrangements were for a Special Purpose Vehicle (SPV) Company called 'Skifeco' to be configured as a cut-out mechanism and to insulate Trio Defence, Alsco, and White's from any direct connections with the Indonesian Government. Banking and capitalisation arrangements of Skifeco were eccentric, seemingly structured to allow for hefty chunks of Arms Cash to be transferred through it to Indonesian intermediaries by Whites and to allow Whites to take a huge commission for themselves. According to the documents, Trio Defence had to capitalise Skifeco to buy its own products, but White's controlled Skifeco and all interfaces with the Indonesian Customer. Weird.

The equipment to be supplied included Simulators and US adapted weapons for training and Skifeco would be collapsed as soon as delivery to Indonesia was completed. Neither Trio Defence, Alsco, or Whites, were smart enough to eliminate all traceability though, as they had to establish a Letter of Credit (LC) for the sale. Their own Banks wouldn't touch it, but HSBC did. The LC was linked to Trio Defence, leaving White's untouched and unmentioned. White's would control the End-User selling price, taking a huge margin for themselves and paying-off the Agents. Trio Defence had no control over anything despite having to capitalise Skifeco and they took all the risk. Trio's

MDs had been well and truly mugged, unless of course they had a personal Arms Cash back-channel.

White's 'man in Jakarta' was an ex UK Defence Company Executive and he managed all in-country aspects of the deal via a lady code-named 'Hairdresser'. Read back a bit and guess who that might have been? I challenged the joint MD's and the US CEO about all this, but the two MDs ran for the hills and wouldn't discuss it, whist Alsco's US CEO said I was being negative and told me to find a work-around.

I double-checked the official UK Export situation and spoke to the US State Department for an Export ruling. Despite the lack of any formal UK Embargo, UK DESO advised that Trio Defence would not get a UK Export Licence because the American Embargo was still in place and they didn't want to upset the Americans. The US State Department cheerily advised an emphatic 'No' when I called them to ask the same question.

By then we (Trio) had much better and cleaner business than this to pursue and I was getting significant success in other territories, such as Saudi Arabia. The twin MDs remained fixated with Indonesia though, and it seemed pretty obvious that there was a kickback agenda, and that I was in the way in my newly acquired mode of 'Poacher turned Gamekeeper'.

The Mother of all indiscreet international 'phone calls added even more spice to the mix, when during an open-line international Conference Call, the US CEO actually said *"...Indonesia is one of the most corrupt countries in the world..."* so, he was not under any illusions about that, but then added *"...we're paying-off half of Indonesia in this deal..."* I winced. I was used to discretion in Defence deals so this was truly awful, and everyone heard him say it. I wrote it down in my notebook and even the two MDs looked really uncomfortable.

I'd gone about as far as I could trying to get this done, either properly or not at all. Many of the 'old ways' of the Defence Industry had changed by then and the Trio Defence MDs had to move with the legal times. They were blundering around blindly in the Hall of Mirrors. I found a clause in the Whites Agreement requiring Trio to obtain a UK export Licence for the deal, and this represented a genuine let-out option for Trio because UK licensing was not possible as I'd discovered. Nonetheless, the US CEO and UK MDs were adamant they would continue with the deal.

I flew out to Singapore on my last trip for Trio Defence to close down Trio Defence's 'loss-making' subsidiary there; it didn't really make a loss, but Alsco had cash-stripped it and left it bare and unable to cover its own running costs. The US CEO wanted it closed-down as soon as possible. It was nostalgic, as previously I'd competed with Singaporean Defence Companies in the Middle East. Were they immune to the lure of Arms Cash? No, of course not.

A man called Eng Heng Chiaw was accused of offering S$500,000 to a 'Defence Science and Technology Agency' Executive for helicopter Tender information. A first version of this charge included reference to aerospace giant EADS but the Company denied any connection with the charge, so it was dropped. The wording was then amended on a subsequent version of the charge to exclude EADS name and this new charge stated simply that the accused made a proposal '...*in relation to his principal's affairs...*' without identifying the Company for whom he was working. He was jailed for just eight weeks, which must have been a considerable relief as a five-year sentence was available.

I met up with Trio's local MD in Singapore to explain why Alsco wanted his subsidiary closed and he and a local business partner bought the Company for a song. It was a good deal for them, and the Company was still in business after both Alsco and Trio Defence went bust. My stay in Singapore was brief, so there was little city-walking to be done, and instead I saw a lot of my hotel room. The 'phone lines glowed white hot from the continued bickering with the US CEO over Indonesia. I had one final trip to make to the USA, so I was in Fort Lauderdale for an Investor Board meeting a few days after my Singapore visit.

I arrived wearily at the Meeting in steamy Florida, where the Investors were excited about Trio's progress. We still butted heads over Indonesia though and I'd made the issues as clear as I could, but even Alsco's own US Government Programmes Executive thought we should ignore the US State Department and go ahead anyway, telling me that '*there are no morals in the Defence business*'. He was talking to the wrong man on that subject. I'd been in the Defence Industry a long time and felt the keen winds of legislative change blow. You have to adapt. But listening to these folks I wondered if the USA really was the home of the 'Foreign Corrupt Practices Act'?

234

By then, the Trio Defence Marketing MD was trying to fix things for Indonesia directly with White's and the US CEO behind my back. They had common cause; Arms Cash. The effort of trying to save the Company legitimately whilst travelling hard and winning new international Orders amidst the constant transatlantic bickering and scheming had taken its toll though, so I left the Company. Maybe they were all just smarter than me, since much later I discovered that Alsco completed the Indonesian deal from the USA with White's and exported the equipment from the USA via Singapore. But Trio Defence went bust soon after and so did Alsco, with one of their funding wizards ending up in a US jail for fraud.

Here was an Arms Cash deal which breached the US Foreign Corrupt Practise Act (because of the payment of illegal commissions), the US State Department Embargo (on Indonesia), probably US ITAR laws also and UK DSO advice, so that's a Royal Flush. Luckily it was on a mini scale by Hall of Mirrors standards, just a few millions, but think about it for a moment, if amateurs like that get away with it then just imagine what the trans-national Defence Corporations can do. In the 'Bad Prospects Hall of Fame' though, there was always competition from Nigeria.

Postcards from Lagos and Makurdi *via a Mews house in West London.* The instruction came down from the MD personally. I was to meet with a contact who was acting as an Agent for a potential Defence deal in Nigeria. It must have been important because I was diverted from my tour of Eastern Europe (this term morphed into 'Central Europe' once we all started to visit regularly) because of the apparent urgency and high value of the potential business. I wondered why the MD had tasked me with this Prospect as he apparently knew the Agent personally, but the answer to that soon became obvious.

Luckily, I didn't need to go to Lagos for the meeting, just as far as a fashionable Mews house in West London to meet this Agent, a Peer's relative, to learn about this 'must-win' Prospect. The MD's Brief was unhelpfully flimsy, but the Prospect apparently related to military aircraft and that was all there was by way of background information.

I didn't embark for the meeting with any enthusiasm, as my previous, also involuntary foray, into Africa had been to Zimbabwe, and that trip was a complete business dud. Africa was way off my regular beat. I had no contacts there and it was largely the province of freelance dealers and brokers, who could easily out-Arms-Cash most regular Companies, except and notably in South Africa, where high-end Corporate Arms Cash incidents swirled and heaved like the angry Southern Ocean.

The first rule in so many, though not all, international Defence Sales situations is to 'follow the money' or, failing that, create deals which are attractive enough to stimulate or attract financing. Africa was unfamiliar territory to me financially, so I didn't know where I'd locate financing for a contractorised Facilities deal say, if required, whereas in the Gulf it was a different story. Locating deal or Facility finance there took as long as it took to dial the right number.

The Company I represented was well-established, well-known, and relatively well-behaved in the Arms Cash sense and had no track record in Africa, and no contacts there either. So, it was with a blend of curiosity and reluctance that I presented myself at the London Mews house where the aspiring Arms Cash Agent was waiting for me. We settled down with drinks but his very first words almost sent me straight home again. He really did want to discuss a used-military aircraft project in Nigeria. I'd hoped that had been a misunderstanding or even a joke.

The very mention of Nigeria to most established Defence or Aerospace Companies at the time, well, those with something to lose anyway, caused rapid retreat. Nigeria was then a Country which I and others like me laboured long and hard to avoid, as it combined a First Division reputation for corruption with a matching reputation for Corporate and personal insecurity. The massive Nigerian 'Armsgate' scandal, cited at c. $2bn worth of Arms Cash, implicated thirty-six military Officers and numerous Government and key business figures in a whole range of Arms Cash activities. As many as two hundred and forty-one organisations or Companies were believed to have received Arms Cash sums from Officials, but at time of writing there have still been no convictions. There has been one 'Armsgate' assassination (so far) so in January 2017, and the Nigerian Government requested the Courts to protect the identities of prosecution witnesses for their own safety.

Nigeria was not on my personal travel bucket list and this book is not big enough to carry all the examples of corruption in-country. A Transparency International (TI) Report observed that corruption had undeniably impacted negatively on Nigeria's security, and a friend of mine who visited the country didn't even reach his Lagos hotel without being mugged by the airport taxi driver. It was just one of those countries which was virtually impossible to navigate in the conventional business sense, without an intimate knowledge of the critical but ever-shifting local touch points, and even then, everything was going to cost Arms Cash and lots of it. Companies trying to stay on the right side of anti-corruption legislation simply didn't go to Nigeria, there was no point.

A plethora of Nigerian Agents and Runners emerged every time there was even a whiff of a Defence Requirement and many approached and tried to hook-up with Western Companies, or simply invented spoof Requirements of their own as money-generating projects. The exception proves the rule of course and I've met folks who've done Defence business in Nigeria, notably some Austrian contacts I had, and years later I had a business near-miss there myself promoting a Special Forces Training Facility, but lost out to Israeli bidders.

The Nigerian corruption element didn't deter every Defence Company though and some had made tons of money way back in the Sonny Abacha era. But the extensive history of rip-offs, the uncertainty and the general fractious state-of-affairs, deterred mainstream players. It was mostly fringe players who operated there, taking their chance on making a quick Buck. Nigeria was firmly off-limits for most 'respectable' (!) Defence and Aerospace Companies, which had other things to do and other places to go.

Nigeria was also one of those countries from which magic 'lists' circulated regularly, rather like the Iranian Lists of requirements during the Iran-Iraq War. I recall extensive lists of arms and ammunition being circulated by a selection of Runners with Nigerian connections, and I hoped this Peer's cousin was not going to push a similar list across his table to me. I'd passed the point of curiosity as to my MD's personal motives over this business whatever it was going to be, in fact I'd progressed directly into the suspicion 'Red Zone'. Get your suspicions in early in this Business if you want to live long and prosper.

Thus, I started off from a sceptical position with the Peer's relative, so it was up to him to convince me that his Project, whatever it was, was viable and achievable without going to jail. I was curious as to what could possibly differentiate his project from so many other Nigerian

Defence 'Prospects' which had crumbled to dust, usually under the unbearable weight of Arms Cash demands, and it didn't take long to discover that he was way out of his depth.

His 'cunning plan' related to the Nigerian Air Force *'Jaguar'* Strike jets, around eighteen of them, which had been sold to Nigeria many, many years previously by the old British Aircraft Corporation, which later morphed into British Aerospace, and then into BAE Systems. The final deliveries of these jets had been made around 1985 but the planes had not been flown much and reputedly did not fly at all after 1990. Nigerian attempts to purchase additional Fighters had been vetoed so consequently there'd been various proposals to refurbish these existing aircraft, but these efforts came to nothing and the planes' decaying status was well known within the Industry.

The proposition outlined to me in the London Mews House was to 'acquire' these aircraft, remove them from Nigeria, then sell them on for use as Spares to another Jaguar-User, specifically the Sultan's Air Force in Oman. The Peer's cousin explained to me that the Company, not him of course, would need to hire a huge Antonov Transport aircraft to fly into Makurdi where the Jaguars were based, and load them discreetly *'...under cover of darkness...'* he advised specifically.

He wanted us to fabricate a cover story about selling the planes for scrap so as not to raise local expectations as to their value, as this might cause trouble, and he was at pains to urge me not to alert the market or any local Players to this 'gilt-edged' opportunity. He thought these old 'planes were worth a fortune, or at least that a fortune could be conjured into the equation. There was of course, to be a generous distribution of Arms Cash to make it all work.

I wondered how he'd been sucked-into all this as he had no background in Defence, Aerospace or Nigeria, and the obvious answer seemed to be his cousin, the Peer. So, in my eyes he was actually a Runner. He offered no detail as to any of the critical touch points in-country, nor any of the operating detail essential to execute such a ridiculous plan. He expected our well-known Company to put together and implement the whole logistic plan for the extraction, movement and subsequent sale of these planes, but offered no answers as to the reality of the Oman buying side of the deal either. I knew a lot about the Oman Air Force then and none of this sounded right to me, no one in Oman would buy into this proposition, it was more akin to a

'B' Movie script than a business proposal. The very concept was farcical.

I returned to the Office and briefed my Main Board Director about the proposal, and he listened with incredulity as I unveiled the details of this 'Prospect', which was after all the MD's pet project. In response to my own Director's question *What do you make of it, Ralph?* I responded, '*Unadulterated, jail-baited fantasy*'. My Director left the Office immediately to corner to the MD, who coincidentally had just returned from Moscow of all places with an even more bizarre Agency Agreement, even by the low standards of those days.

This new Agreement, to market what were described as 'certain products', fell a long way short of the above-board Defence and Military aerospace field in which our Company was active, and a copy of the document was passed discretely to me for review and analysis. The very first item I could discern, despite the language issue, was 'T80'. I know a Russian Main Battle Tank designation when I see one and the list was packed with Arms and ammunition types. The MD left the Company shortly afterwards. The practise of individuals 'freelancing' inside a branded Defence or Aerospace Company has always been a risky undertaking but it's not uncommon. The Nigerian Jaguar Project which never-was, was removed easily from our radar, but coincidentally, I was approached soon afterwards by another Runner with a different Nigerian proposition. How was I attracting all this Nigerian stuff?

This quirky Runner-cum-fantasist (there are so many of them in this Industry) was a former UK Civil Servant who'd morphed into a Runner for a mysterious Irish-based Broker. He proposed an armoured vehicle refurbishment programme in Nigeria but needed a sponsor Company to support the venture. The vehicles in question were the Alvis manufactured light tanks, CVR(T) types ('*Scorpions*' etc.) and several smaller Companies might at least have entertained the proposition. But once again, the very mention of Nigeria in those days tainted the prospect and there were few, if any, takers amongst reputable Companies.

But it was the next part of his pitch which ended our conversation abruptly. He claimed that although he'd left the Government Vehicle Repair Agency where he'd worked some time ago, he remained '…discretely in Government employ…' in what he called 'another capacity'. This is another familiar Runner approach. The pitch is designed to delude a Company into believing that the individual is connected in some way to the Intelligence or Security Services, and that he thus has special capabilities, special powers almost. This was nonsense of course and again we declined to get involved, but there was a curious footnote to this episode. The same individual re-emerged years later, still in connection with Nigeria, but this time he was the Runner in a plan to export Ukrainian Defence materiel. He went to Ukraine to arrange the details, but the deal apparently soured when someone reneged on the Arms Cash element of the deal, which was not a very smart thing to do in Ukraine.

I learned later on the grapevine about the Ukrainian aircraft munitions-carrying incident in Nigeria, and later still, that a list of British Defence brokers had been passed to the UK Government (directly to the PM it was claimed) by a Ukrainian Deputy Minister for action. Was it all connected? Who knows, who cares. There was better and cleaner business to do, and most of the Corporate Defence Company World had modernised and moved on from this sort of activity. Defence Industry Junk-Bonds.

Defence Companies which had been willing to make Arms Cash deals in the past, had undergone legally-prompted Damascene conversions and become risk-averse Angels. But these incidents stand as shining examples of the sort of peculiar approaches which passed across an international desk in a Defence Company. You just had to accept that Walter Mitty had Hillbilly Cousins and that you were destined to meet most of them in your Office, on a plane, or in some foreign hotel Lobby. Just don't get sucked in if you want to keep breathing and stay in a job.

In more recent times, Nigeria has made giant strides in cleaning-up Arms Cash issues to the extent that even the Press there campaigns on the issue now. The Nigerian Guardian commented on Arms Cash issues thus:

'Corruption such as procurement corruption is responsible for the buying of substandard weapons thereby exposing the fighters to the dangers of being constantly overpowered and killed by these terrorists. The danger in permitting the existence of corruption in the military is that the institution of the military or the armed forces will be inhibited from actualising the mandate of protecting the territorial integrity of Nigeria.'

Maybe they've entered a brave new world now, who knows. With the Nigerian ghost Prospects exorcised, I hoped my African adventures were over, but sadly, they were only just beginning, and I was despatched further south.

Postcard from Harare, Zimbabwe *following lunch in Milan.*
I still don't know how I got mugged into accompanying a US Aircraft manufacturer's Executive to Harare, and frankly, I was surprised that any major American Military Aerospace Company would want to go there, or even be allowed to visit during the Mugabe era. All I did know was that I didn't want to go there.

I met my US contact for an initial Briefing and a lovely lunch in Milan, and a few days later off I went from Heathrow via Air Zimbabwe, or whatever it was called then, on one of the worst and most unpleasant commercial flights I've ever taken, even pushing my Ukrainian Antonov nightmare into second place. My Contact and I travelled independently and just before the Arms embargo imposed by the UN, and implemented in UK law, came into force.

Zimbabwe was a highly controversial place for any Defence Executive to be seen in, and not just because of all the Human Rights issues and other domestic excesses associated with Mugabe's autocracy. Arms Cash issues there were never far away. In 1998, Zimbabwe Defence Industries (ZDI) sold ammunition and uniforms to Angola, violating the Arms embargo set-up under the Lusaka Protocol of 1994, which attempted to heal the ongoing sore which the Angolan conflict represented. ZDI was also recorded as having *'irresponsibly exported'* fifty-three tonnes of ammunition, and there were incidents of supplies of ammunition to Mercenaries in Harare also. The Zimbabwe Government tried to buy Arms and Ammunition from China during the Embargo period, so generally speaking, the Defence Business had been pretty-well fizzing in Zimbabwe one way or another.

My new friend was an optimist who hoped and indeed expected, to sell Training aircraft in Zimbabwe. But what guarantees would or could be available to ensure Training use only for such Aircraft, even if the Government could pay for them? Training aircraft could be adapted easily into combat or counter-insurgency roles, as demonstrated in the stories circulating about British-supplied Hawk Trainers being used operationally by the Mugabe Government. But Politics and embargoes aside, I was surprised that this major US Company needed such small-fry business. Could they even find Zimbabwe on their Marketing map? Surely not. Were they up to speed with the Continent's Arms Cash challenges, did they even care?

A veritable ocean of Arms Cash issues ebbed and flowed across the headlines in the wider African region. Defence corruption charges were laid against Jacob Zuma when he was South African Vice-President but were later withdrawn owing to lack of evidence. His former financial Adviser, Shabir Shaik, was convicted of corruption and fraud in a case which featured an infamous encrypted Fax (it's those faxes again) referring to a meeting between Shaik, Zuma, and Alain Thetard, then head of French Defence Company Thomson-CSF (now Thales) in South Africa. But Shaik's was the only subsequent conviction at the time, and although he was released early from a fifteen-year sentence on medical grounds, he was soon re-arrested. Charges against Zuma also ebbed and flowed.

It wasn't just Thales who wanted South African Arms business though, and it was alleged that BAE Systems paid millions, c. £100m according to Press reports, in secret commissions or incentives to win a contract to supply Hawk/Gripen jets to South Africa. Michael Woerfel, the German former head of EADS in South Africa, also faced corruption and fraud charges which were later withdrawn. Tony Yengeni, once the Chief Whip for the ANC in South Africa's Parliament, allegedly promised to use his influence to promote a bid for an Arms contract by EADS, and subsequently served five months of a four-year jail term for defrauding Parliament in connection with an Arms contract.

Various major Defence Companies, including BAE, SAAB, Augusta, and Thales, all fell under scrutiny in South Africa following corruption allegations relating to the massive South African programme to acquire new fighter jets, helicopters, submarines, and warships, valued in total at around £3.9bn. Quite who the South Africans were planning to fight with all this expensive hardware remains a mystery. Maybe it was really for onward transmission, who knows? A national Commission of Enquiry subsequently exposed

Procurement failings in respect of these various Arms contracts (shock!) but don't hold your breath.

Elsewhere on the African Continent, BAE had settled with the UK Serious Fraud Office over business activities in Tanzania, and the Company paid out a record £30m penalty for *'failing to keep reasonably accurate accounting records'* following the highly controversial sale of Radar there in 1999. BAE Chairman Dick Olver admitted that BAE had made *'…commission payments to a marketing adviser and failed to accurately record such payments in its accounting records.'* How do you even do that? How do you fail to accurately record the payment of millions and millions of pounds? Didn't anyone notice the cash flowing out? It reminded me of my own experiences in the Gulf with the stolen then repaid Arms Cash commissions. Just how does a major UK Plc., or any other international Company for that matter, account for these Arms Cash transactions? How do Auditors miss millions of 'lost' illicit Commissions?

South Africa's former indigenous armaments agency, Armscor, was famously awash with widely-reported corruption allegations, but in fairness to South Africa, the 'Directorate for Priority Crime' (known as 'The Hawks') is currently investigating a variety of fraudulent Defence deals, including one concerning the provision of PPE during the COVID outbreak, which puts it well ahead of the UK.

Nonetheless, my US Aerospace friend and I did fly to Zimbabwe to do military Trainer aircraft business, and yes, we did visit during Mugabe's reign and despite Zimbabwe's controversial use of military training aircraft on operations elsewhere in the region. I don't mind repeating that it surprised me that the Americans were happy to even try and trade there as they surely had better business prospects to pursue elsewhere. Clearly, it was not just Albion which was perfidious.

My friend assured me that he was about to clinch a deal and I was briefed to piggyback on this deal to enhance it with a support and maintenance package. The deeply flawed Executive Board logic at home was that historic UK connections with Zimbabwe (!) would ensure a better ongoing support relationship than a direct US one. In reality the US Company just wanted to sell their planes and run off, leaving us to do the grubby support bits of the job and we were gormless enough to fall for it.

I stayed in the legendary Meikles Hotel in Harare, where two magnificent, though slightly sad and careworn Lion statues stand guard. Meikles had sustained its premier Hotel status in Harare since 1915 but that was probably not too difficult to achieve given the state of their competitors. The hotel was, as a passing Lobby Hooker explained to me as I was checking in, also a monument to Colonialism.

From Meikles we made our sales and marketing forays out into the Defence and Air Force Departments, though we were strongly advised not to go wandering around unescorted in Harare. The streets around Meikles bristled with street-sellers peddling carved wooden African animals, Batique prints, and similar folk-crafted items, but the smack of poverty and oppression everywhere was unmistakable. The brief pedestrian excursions I did make revealed a City in sad and ongoing decline, with an atmosphere that felt heavy and menacing for visitors; it was no place for an extended city walk. In May 2006, an editorial in the *Zimbabwean Financial Gazette* described Harare as a '*sunshine-city-turned-sewage-farm*' and at the time of my visit I could see and smell this clearly enough.

My optimistic Aerospace friend remained confident that he would sign a deal and showed me the mountain of paperwork his US Company burdened him with in order to appoint a local managing Agent. Driven by the US Foreign Corrupt Practises Act (FCPA) he was compelled to complete a virtual book of forms in order to make a 'clean' and acceptable Agent appointment with no tangible links to Rulers, Ministers, Senior Officials, Officers etc. It took forever to get all of this done. How did they ever expect to win anything? Since that visit long ago the UK Government has enacted the 'UK Bribery Act' so even the humblest overseas local Agent or Representative is now viewed with great suspicion even if they are not actually Arms Cash recipients; indeed, the very word 'Agent' has become virtually taboo. Defence Industry Woke.

By the end of the trip and predictably, no progress had been achieved on this feckless Sales mission, and my travelling aircraft salesman friend was crestfallen to be told finally that the Air Force would not be buying any of his expensive aeroplanes as they had other plans. They later bought Chinese aircraft. I'd had enough, so I took a Zimbabwe Airlines flight back to UK, the second worst long-haul flight of my Career after the outbound Zimbabwe Airlines flight.

Imagine though the furore if his US-manufactured Trainer aircraft had been bought and then adapted for and operated in offensive operations in that region. I hate to fail, but this was a good deal not to get, and I was convinced it never really existed as a serious proposition in the first place. It was just another unpleasant and personally risky trip to another unstable country for nothing, and probably represented no more than a tick-box exercise in the Boardroom at home. I can just hear it now, some thrusting Main Board Director sounding-off to his colleagues 'Yes, we've working on Africa…we have someone in Zimbabwe this week.' Sometimes you just can't win. Sometimes you don't want to win. Just occasionally you get to go somewhere really nice.

Postcards from Rome and Milan

I flew to Rome to meet some new business friends and after staying overnight in a wonderful family-owned Hotel, I was driven into the hilly outskirts of the City. The car stopped outside the lovely pine-shielded Period building which is home to the Guardia di Finanza. It was the Guardia di Finanza which broke the Arms smuggling ring that shipped Italian military products to Romania to mask their final illicit destination, Iran.

The Guardia's attractive building had been the Italian Track and Field Team's base prior to the 1936 Olympics, and this architectural heritage was reflected decoratively inside the building. Colourful glass insets in the doors portrayed Art-Deco images of athletes and the old cinder running track still existed, in part, when I visited. Sadly though, this had been slightly truncated by an expansion of the Guardia's Facilities, which included the Shooting Range which I was there to visit.

I watched Officers practising against a variety of targets and as we left the site, I commented to my Host on the attractiveness of the location. He agreed it was a delightful spot but advised against visiting too frequently. These Officers had to recover Mafia money and the place was sometimes watched he said, it was neither smart nor healthy to be seen around here too often. Business introductions aside, my new Roman friends also introduced me to new business jargon, when after presentations to one potential Customer one of them told me…

"Ralph, we have flirted with them and they are ready; now we have to f*** them." I didn't relay that advice to the Board at home. There was no point, they were already f***ed.

I've indulged myself thoroughly with lengthy City walks on each and every subsequent Business visit to the Eternal City, and on one particular trip, when I stayed close to the Coliseum, I wondered if the Roman Army Buyers who procured the Gladius etc. for the Legions, were recipients of Arms Cash kickbacks paid in Denarii.

My later visits to Rome included meetings with the Italian MoD to discuss Sea Mine clearance in Coastal areas which were designated for development, and these discussions emphasised the enduring hazards of legacy Defence products. The Coastal waters still contained unexploded Allied Munitions from WWII. Decades after the end of WWII, the German bombing of London still presents unexploded Ordnance issues as does the UK bombing of German cities. The Middle East will endure long-term issues as a result of Gulf War(s) Mining and from Combat throughout Kuwait, Iraq, Iran, Syria generally.

These days though, the Victors also do get the opportunity to profit from the damage and destruction they caused in the first place. Clearing Kuwait of Mines for example resulted in the world's biggest commercial Mine Clearance contract, which provided a great example of profiteering from Conflict. The UK AEA even proposed a chargeable service to survey and clear the residual toxicity in Kuwait caused by Allied Depleted Uranium munitions and their helpful Report spelled out the problems thus:

'Handling heavy metal munitions does pose some potential hazards, as does the possibility of the spread of radioactive and toxic contaminations as a result of firing in battle… and can become a long-term problem if not dealt with …and [pose] a risk to both military and civilian population.'

The Report indicated that the tank ammunition fired by British and U.S. during the Gulf War amounted to 50,000 pounds weight of depleted Uranium and further stated that…

'...If that amount of DU was inhaled it could kill 500,000 people...' but the report added *'...obviously this theoretical figure is not realistic; however, it does indicate the significant problem.'*

So reassuring.

I encountered no Arms Cash requests on my Italian trips although the Italian Defence Industry was no stranger to the topic. Italian Defence Company Finmeccanica fell foul of regulations in a big way in India when their former CEO, Sgr. Orsi, together with the head of a subsidiary Company, were accused of corruption and false accounting. They were found guilty and sentenced, only for the sentences to be overturned when a Court in Milan concluded that the facts of the case were insufficiently proven. The Indian Government, however, cancelled their Helicopter contract worth c.$660m million which was at the heart of the allegations.

In another case and according to *ProPublica*, it turned out that Italian Sniper Scopes had made their way into Taliban hands in Afghanistan in 2008, probably as an outcome of the activities of the so-called 'Iranian Arms Ring' which operated in Milan to circumvent the international Arms embargo. That story reminded me of my Iranian contact experiences in Dubai years previously, with echoes of my encounters with the sultry femme fatale Fatima. The Italian operator of the Milan ring was overheard on a wiretap to complain that he had to make regular payments to a local Politician, Arms Cash kickbacks.

According to ProPublica, Milan was at the centre of this conspiracy involving Italian Arms Dealers and Iranian Runners, to sell a whole variety of Arms Cash-earning items including ammunition and helicopters to Iran. Some items, like the Sniper scopes, evaded the embargo completely and were used subsequently against Coalition Forces in Afghanistan. Nine Italians and Iranians were arrested in connection with this, but Iranian buyers and Runners remained actively on the look-out for other Arms Cash opportunities. My initial trip to Milan was vanilla by comparison to all this, as I simply met up for a good lunch with a US contact connected to an African Trainer Aircraft project. I was free to enjoy the food and the Milanese fashion, without having to watch my back all the time. Well, not too much anyway.

Postcards from Islamabad, Karachi and Rawalpindi

My first Defence Business trip to Pakistan was a challenge even before I landed. In fact, it was a challenge because I couldn't land. The aircraft was tossed around the sky like a fluttering kite during a violent electrical storm which lit up the night sky with brilliant silver flashes, and which created nausea-inducing clear air turbulence with heart-stopping abrupt drops in altitude. Passengers looked afraid and some were actually praying as the plane plummeted violently.

The Pilot made two valiant attempts to land, pulling out at the last possible second on both attempts, and after what seemed like an eternity, he abandoned the effort and headed off instead to Karachi. Passengers were disembarked in the early hours of the morning into a closed, dark airport where they had to wait and wait and wait in an un-airconditioned, zero-services concourse with no Staff at work. Dishevelled passengers curled-up on Terminal seats or lay on the floor and others sat and just cried. We looked like an Asian version of the Retreat from Moscow.

The following morning, after a seemingly interminable wait, we embarked on a different plane and this time landed at the first attempt in Islamabad, where the airport terminal was even more chaotic than Indira Gandhi airport had been during my first trip to India. I spent a few days in Islamabad for meetings before heading south to Rawalpindi and the Pearl Continental Hotel, but barely a day into my Rawalpindi stay I was struck down with a severe illness for several days. It's shocking to realise how quickly you weaken, and at one point I hobbled out weakly to find a pharmacist and only just made it back again before collapsing.

Once recovered sufficiently I set off to visit the Pakistan Ordnance Factory offices in the Wah Cantonment area. Many Victorian Raj military terms like 'Cantonment' survive in Pakistan; remember the bin Laden '*Abbottabad*' raid? That nearby Afghan town was named after a Victorian Army Major, James Abbott. My own loose personal military connection to Rawalpindi stemmed from my ex-wife's great-grandfather, who had been a Sergeant in the 4th Dragoons, and who died in Rawalpindi in 1901. I found a memorial plaque bearing his name in the old garrison church, Christ Church, which I persuaded someone to open for me. It was immaculate inside, but the cemetery was a jungle.

It was higher-ranking ex-Army officers than the late Dragoons Sergeant who were my company's legitimate representatives in Rawalpindi, for as an established Supplier, the company was allowed official local representation to manage contractual matters. This modest local level of permissible representation contrasted massively with the very high-level covert Arms Cash scandal which enveloped the Franco-Pakistani Agosta Submarine deal during the 1990s 'reign' of President Benazir Bhutto, when multi-millions of dollars were secretly paid out in commissions to fixers and middlemen, in the so-called *Karachi Affair*, and there have been other Arms Cash cases since in Pakistan.

My company was fortunate to have an excellent ex-Brigadier and his retired Colonel sidekick on board as they were keen businessmen, good company, and diligent without peddling any Arms Cash agenda. They genuinely wanted the best they could get for the Army and I had great respect for them. The Colonel's son was a serving Army Officer and was deployed at the time in an operational position high on some Kashmiri mountain, facing out across the barren slopes toward his enemies in India.

During convivial lunches in the clear-aired Murree Hills high above Rawalpindi the Colonel related some of his own experiences of the previous conflicts with India. But when I proposed to visit Peshawar on a forthcoming free weekend, the Colonel advised me strongly against it, and I deferred my excursion on this advice. Later, I heard that two Westerners had been killed there by a marketplace bomb explosion that weekend, and this was well before the ill-advised Western invasion of Afghanistan disturbed everything else. It is widely forgotten now that way back in 2002, a young US journalist, Daniel Pearl, was abducted and beheaded. Sadly, I cannot venture back safely to Pakistan or to many of the places I visited so freely and so confidently in those days, which is a terrible shame. Such a beautiful country, such repressed potential, such National pride.

It was the potential for abrupt political change which occupied my mind on a later trip into Islamabad, when Military Command and Control Systems were on my agenda. This time my local agent was necessarily a more political animal than my ex-Pakistan Army contacts in Rawalpindi had been, since C4I Systems were very expensive and needed a much higher level of 'input'.

My new 'friend' Ali was young and cocky and walked with a swagger. Like so many young Pakistani business professionals his business interests aligned with a genuine appetite for Nation-building. I liked him. He was bold and brash and viewed problems straightforwardly, simply as challenges to be overcome, and he was relentlessly optimistic about the future of Pakistan. I truly hope he still is. He also had one great social advantage, for he was the son of a retired General who also served in Parliament. Pakistan is a famously difficult country to govern and has 'enjoyed' over thirty years of military rule since its foundation in 1947.

One evening Ali took me back for supper to the big family house where I met his father, who was every inch the upright professional Pakistani military man, full of integrity and high standards. After a meal awash with geo-political chat but without wine of course, Ali and I adjourned to another Reception Room to discuss our business plans, as the General marched off to his Study with his staff and associates. As the evening melted into night, I became aware of a succession of smart-looking cars arriving at and departing from the house. Each arrival was greeted by a servant who rushed down the steps to hold open car doors for the guests. These arrivals and departures continued well into the early hours, and when Ali and I finished our planning I commented to him…

"Your father works long hard hours with all these early-morning visitors…" to which Ali replied…

"Yes, they're deciding whether or not to mount a coup…"

There can be few phrases the traveller wants to hear less in a country, especially in Pakistan, than that. A coup. At that moment all I wanted to do was head for the airport at top speed and fly out, although airports are not necessarily the safest places to go, as I was reminded during a trip to Turkey.

Postcards from Ankara and Istanbul

I was in a meeting with two Turkish Aerospace Executives in Ankara when one of them was called away to take an urgent 'phone call. He returned to advise me that there had just been a bomb explosion at Ataturk airport and that forty-five people had died in a shooting and suicide bomb attack. I wasn't travelling that evening, as I was due to leave the next day, but the incident was a stark reminder, as if any were needed, that we live in dangerous times. We are all potential targets now and in any international business role you simply increase your statistical probability of being in the wrong place at the wrong time.

I'd visited Turkey several times since I'd started out in the Defence Industry, but after the failed Coup in 2016 many Companies reduced their marketing efforts there, only to rebuild them for the new Fighter programme. Nevertheless, I've always enjoyed Turkey and Istanbul remains my pre-eminent favourite city to visit, with the Blue Mosque and the Hagia Sophia being my favourite of buildings in the world. Anyone who remains unmoved on visiting the Hagia Sophia has no soul. Istanbul fizzes and offers delicious splashes of intrigue, in fact, Istanbul could have been designed for intrigue.

A delayed inward flight on one visit caused me to arrive later than planned at the Istanbul Radisson Blu Hotel, and I was due to meet a private Company Owner with Defence and Aerospace interests later in the evening. The dominant Turkish Companies in these sectors are Government-owned and are well known, so everyone including me meets with Havelsan, Aselsan, TAI and MKEK and so on, hence, a private Company meeting in Turkey was a refreshingly different proposition. I was checking into the hotel late, when someone messaged me from an unrecognized number. It was a great message.

"Go to the jetty behind the hotel at six o'clock..." the message urged. I unpacked quickly, showered and went down to find the jetty, walking through the smart ground floor restaurant, part indoors part al fresco, to reach it; the timber jetty projected into the Bosporus. I was taking-in the magnificent early evening views when an elegant motor launch approached the jetty and eased alongside. It was a gorgeous slinky boat with classic Thirties lines; all mahogany and high-gloss two-tone yacht varnish.

Positioned on the canopied aft deck was a table with four seated, black-suited, sunglass-wearing gentlemen sipping wine. The Turkish Blues Brothers surely? The boat bubbled closer to the jetty, reversed thrust gently, and someone called out 'Jump on Ralph...' Now there's an invitation. Yes or no? I stepped off the jetty onto the boat to meet my new friend and contact, Mehmet, who waved his arm towards the open water.

"Tonight, we eat in Asia" he said theatrically, and with a nudge of the throttle the boat's bow lifted, and we surged elegantly across the Bosporus, finally mooring-up for Dinner at a high-end waterfront property with a huge projecting viewing deck on the opposite shore. My French employers had been sceptical about getting any positive outcomes from this visit, but I got a lot out of it, both personally and professionally. They were in deep shock when an order arrived.

As with so many other commercial situations you soon realise that you are on the wrong side of the table, and would prefer to be working for or with, the person you've come to meet. There was no Arms Cash in this deal, it was all very clean and above board I'm pleased to say, although Turkey was one of the countries listed when Airbus was fined a breath-taking world record amount of £3bn for bribery in around twenty different countries. I've never had any personal experience of any Arms Cash deals in Turkey, although a major UK-based Defence Company I know is still paying a Turkish Company a regular 'Retainer' plus a bonus on contracts, which is simply Arms Cash by another name. But that's small beer.

Serious Defence corruption issues did arise over the construction of a new Turkish Military base in Qatar though. Qatar is Turkey's new political and economic friend and Qatari Intelligence was implicated in offering a huge bribe to fast-track the Base construction Project through the Turkish Parliament. Stockholm-based Research organisation *Nordic Monitor* revealed the existence of an Intelligence Report on the Project and associated bribery, which was submitted as an evidence item in a Turkish court by Rear Admiral Sinan Surer, the former Turkish military chief for external Intelligence. Qatar is Turkey's third largest arms Importer with sales of $180.5m in 2021.

Turkey has sophisticated, well-developed and progressive indigenous Defence and Aerospace capabilities, which are invariably underestimated by UK-based competitors, hence many optimistic Marketing visits to Turkey fail because the targeting homework has not been done correctly. The same applies in India. That aside, my new friend Mehmet turned out to be a great City Host and he introduced me to many places I'd never seen on previous visits, including the Constantine Cistern, that great subterranean Byzantine Reservoir with columns rising from placid waters, and spooky white fish swimming around in the half-light. I must return to Istanbul if only to revisit the Hagia Sophia; the only place in which I can recharge my vacant spirituality. Unfinished business. I concluded business of a different kind across the border though, in Turkey's great regional rival, Greece. I had a long overdue appointment there with an old friend.

A Postcard from Athens

I'd worked for a small Defence Facilities Company which won a Project in Athens for a Police/SF Training Facility, and although I'd never managed to get back and see it finished I remained eager to make at least a fleeting visit to Athens to meet up with an old friend there, a true bandit from the 'Hall of Mirrors'. Zaroubi.

'Zaroubi the Greek' was a long-retired old-style, though small-scale, Arms Dealer who I'd met for the first time in Dubai. He'd initially assumed, wrongly of course, that I was in the Intelligence Services, but I overcame this misunderstanding over a very good Dinner in some expensive Dubai Hotel, I forget which one, they now all look the same inside. We got on well, consumed a lot of booze and I promised to meet up with him on his home turf in Athens one day, because he was unsure as to whether or not he'd ever be allowed back into Dubai. I don't know why, safer not to know probably. I deliberately re-routed a later and unrelated trip to give me the opportunity to fulfil my promise. It was a bit of an indulgence, but I cut off from my schedule and flew to Athens to dine with Zaroubi once again.

Zaroubi was a larger-than-life character with larger-than-life appetites and it was a challenge just keeping pace with him. He alluded to all sorts of secret deals in which he'd been involved over the years, though I'm not sure how much of this was true and how much was bravado. It didn't really matter though, Zaroubi could have come straight from an old adventure Movie as a comic-villain-turned-good-guy, and he embodied everything that scares the pants off modern Corporate Defence Industry Executives. Hooray!

Judged from a Corporate perspective, I'd wasted a day and some expenses, but a hotel night and a great Dinner, did I care? No. I enjoyed Zaroubi's company and we shared contacts. The Company 'owed me' at least one anyway. Zaroubi is dead now, of natural causes I'm assured, so the mould is finally broken and there can there be no more like him. None of the new breed of Eastern European Wheeler-Dealers possesses an ounce of Zaroubi's charm, nor any of his raconteur's skill.

Winning major Defence contracts in cash-strapped Greece was an unlikely outcome at that time though, as Arms Cash deals had impacted adversely on the parlous state of the Economy. Defence

Minister Akis Tsochatzopoulos was jailed for twenty years for extracting €45-50m in bribes for missiles and submarine purchases, and in March 2019, eight people, including senior Defence Ministry Officials, were sentenced for accepting bribes and money laundering. A special Committee in the Greek Parliament was examining a whole series of subsequent deals, involving billions of Euros and other Ministers and Officials for Arms Cash deals, relating to big-ticket purchases of Tanks, Helicopters and Naval vessels.

Zaroubi was not in this Premier league thank goodness, but he never ended up in jail either. I bet he sold Gabriel a Kalashnikov when he reached the Pearly Gates. I hope so. I'd fondly imagined that small countries with equally small Defence Forces were less likely to be susceptible to the Arms Cash phenomenon, but a visit to Slovenia proved me wrong on that count.

A Postcard from Lubliana, Slovenia

Slovenia boasts a population of just over two million people and it's a pretty country with total Armed Forces of barely 6,500 personnel. One might hope, expect even, that such a small and recently re-emerged country would be free of both Arms Cash history and rampant Western Defence Company Sales Executives; well, wrong on both counts. With regard to Arms Cash history, the former Prime Minister Janez Jansa was sentenced to two years in jail for soliciting bribes as part of a Defence deal, after he and others allegedly sought Arms Cash of around two million Euros from 'Patria', the Finnish Defence Company (again!) for the supply of Personnel Carriers in 2006, a deal which was subsequently scrapped... and rampant Western Defence Sales Executives?

Well, I arrived shortly after the 'Velvet Revolution' with a very modest Defence Aerospace agenda which progressed nowhere, but this mattered not a jot, to me. I'd never been to Slovenia before and my visit to fairy-tale 'Bled Castle' coupled with city walks around Lubliana, the city of the three bridges (yes, even more of those European city bridges) made my whole trip worthwhile. I pushed without enthusiasm at a couple of Prospects there, but never made it back to Lubliana on Defence Business. I did go there again but for an ICT Company a couple of times, and my local friends told me it's now 'Russia's Switzerland'. Maybe a good place to retire to, more so than Albania that's for sure! Or maybe not...

Postcards from Tirana and Durres

I arrived in Albania following the most appetising Brief I've ever been given in the whole of my international career. The CEO of the multi-faceted privately-owned ICT Company for which I was working grabbed me as I passed his office.

"...apparently, we've got some friends in Albania, Ralph. Get over there and get me some business..."

I boarded a 'plane two days later and fetched-up in Tirana, imbued with all the popular images of Albania which prevailed then and which still do in some quarters, i.e. it was a land of Gangs and Gangsters, a place to be avoided. My Contact met me at the airport and although I really wanted to get to the Intercon' (yes, another one) to shower after the flight, and to prepare for my meeting with 'who-knows-who' the following day, I was driven instead to a city centre Restaurant for a massive folksy meal washed down with gallons of Raki. I finally made it to my hotel bed late and feeling awful and was advised by my Contact to be ready for a meeting at eight o'clock in the morning.

I was collected bleary-eyed from the Hotel in the inevitable black Mercedes and whisked-off to a scruffy Soviet-style paint-peeling building, with Armed Guards and barbed wire outside and with numerous layers of security inside. We ascended several flights of concrete stairs, and the décor seemed to improve with each new level. I was still in the dark as to who I was going to meet and was deposited in the office of a very attractive and super-savvy US-educated Executive Assistant. There was a long pause for coffee and then more coffee, until finally she announced

"...the Prime Minister will be with you shortly..."   and I was ushered into an impressive Conference room with an aircraft-carrier-flight-deck-sized conference table to wait for him. The Prime Minister hadn't been on my list of likely meetings, so I switched the Presentation format I was due to make to a high-level one. The Prime Minister listened attentively; he was courteous, he asked good questions and best of all he promised that his Executive Assistant would arrange Ministerial meetings for me, and so she did.

The Defence component of the Company Offer was only one part of the varied Portfolio available, so I was delighted to meet a selection of Government Departments which increased the possibility of getting something out of this visit. After meeting various Ministers as

255

promised, I finally met the military Officer who headed up Defence procurement activities.

Albania had either been persuaded or volunteered to contribute a small military contingent to the Coalition military effort in Afghanistan, so for the very first time Albanian Forces came face-to-face with the Military technology which western military forces take for granted. As the good Colonel put it to me '...we had absolutely no idea how far behind we were...' No doubt that was true, although the vast wealth and hi-tech military inventories of the Coalition had still not suppressed the Taliban.

Away from official meetings, I explored bits of Tirana and found a small cemetery with a Memorial to the fifty-three British servicemen who died in Albania during World War II. Albania had been of great interest to the British Special Operations Executive (SOE) during WWII and to Western Intelligence communities too during the subsequent Soviet era, so shades there of my early MoD indoctrination with the 'Eastern Approaches' security training film. Tirana was knee-deep in US diplomats during my visit, which coincided with important Mayoral elections in Tirana.

I also went south to Durres, where superb Roman ruins were begging to be explored, and a whole Tourist Industry was waiting to explode into life; the Italians were already there of course, lapping it up. I didn't get any Defence business done in Tirana, but did do some much more environmentally friendly business on subsequent visits, and in a Country which boasts more Soviet-style concrete border installations than any other, that seemed appropriate. It was a refreshing world-away from the manoeuvring which characterised Defence business elsewhere, especially in Paris.

Postcard from Paris

I long for Paris to be romantic and exciting, somewhere special, but it isn't any more. The widespread aroma of dog-pee puts a damper on things. Years ago, as a Reserve Forces soldier, I competed in a military Endurance event around the course of the River Marne in Paris, and my sidekick and I mistook a Legion Étranger recruiting caravan in Fort Vincennes for the Event Registration centre, which led to interesting exchanges with the Hosts. The overnight accommodation on offer on that occasion was a truly sub-standard stable, so we opted instead to improvise a tactical bivouac in a small wood in a Paris city centre ornamental park. Rain fell upon us in torrents all night.

Those were distant memories when years later I attended a series of Eurosatory Exhibitions, the biennial and giant-sized international Defence and Security Show, along with several Paris Air Shows. Like most Defence Industry Executives I've attended both Shows numerous times, either with an Exhibiting Company team, or as a Freelancer taking appointments, and as with all international Shows the national Hosts get to control visiting international Delegations and shepherd them around. I've also delighted in leaving both these events early, heading off instead to city-walk in the Centre of Paris.

The French so dominate visiting overseas Delegations that it's pointless expecting to see any of them unless you've made prior direct arrangements, and one major UK-based Company even pulled out of Eurosatory once in a hissy fit over it. But hey, it's their Show and we do the same when it's UK's turn. Huge chunks of the UK Defence Industry are in French ownership anyway, so now it's rather like visiting Head Office. Still, getting one's business done and quitting the Show early, ideally at the end of the first day, is the definitely way to go, so time can be spent wandering around Paris looking at more interesting stuff like Art and Restaurant menus, which brings me to another reason for including a postcard from Paris here.

I visited Thales HQ on the edge of Paris once to discuss a Middle Eastern Prospect which I'd advised them to quit, because the Company had no developed Systems to offer and they were being used unwittingly as a stalking horse by the Customer. It happens quite often. The French Marketing Director found it hard to accept they could not compete, and it took a wonderful three-hour lunch to discuss the situation with him. I must congratulate Thales here on still lunching in the 'old style', since many British-based Companies now think they are Americans so don't 'do that stuff' anymore.

Of course, Thales and Arms Cash are old friends and it's hard to know where to start talking about this, since in the words of one US Combat Pilot, this subject is a 'target-rich environment'. Let's start with a couple of statements from the Thales website (as at time of writing):

'Thales introduced a corruption risk prevention policy in the late 1990s in order to protect itself from corruption and bribery, which represent a major risk for multinationals.' and…

'Thales's Corruption Risk Prevention Policy is backed by a dedicated governance structure, including a standalone "Compliance" function with 18 Chief Compliance Officers and over 100 Compliance Officers.'

That's good then, they seem to have Compliance all buttoned up. Except that in April 2023, '*Sputnik Africa*' reported that former South African President Jacob Zuma and Thales were facing 18 charges related to corruption, money laundering and tax evasion, relating to their involvement in a 1999 $2.5 billion arms deal. In 2002 Thales confirmed a media report that the company was being formally investigated in a corruption probe tied to the sale of submarines to Malaysia. It was also reported in the New York Times in 2005 that former Chief Executive of Thales Engineering & Consulting, Michel Josserand, said during an interview with Le Monde, that the paying of bribes by Thales was widespread, saying...

'I estimate that Thales must pay out between 1 percent and 2 percent of its global revenue in illegal commissions...'

In 2023 articles in various Newspapers and websites described how Thales is facing a bribery probe by French prosecutors into a $2.5 billion contract for the modernisation of the Indian Air Force's fleet of Mirage 2000 planes, allegations which Thales denies, and back in 1991 an international court ordered France and Thales to pay $830m in compensation to Taiwan over a sale of warships in 1991, and so on and on. On y va! Nothing to see here.

Postcard from Berlin

When I first joined the MoD as a junior clerk, I also served in a Reserve Forces Infantry Regiment which trained annually in West Germany. The Cold War glowed white hot then and the Soviet Union was Britain's clear and deadly enemy, having been our ally and saviour in WWII. Let's be clear about that, without the supreme Soviet effort on the Eastern Front, there could have been no Allied Victory in WWII.

During the ensuing Cold War years, Britain and the wartime Allies, maintained a huge post-WWII military presence on West German soil and waited for the allegedly inevitable Soviet invasion to be launched. Allied Forces were committed then to the 'first use' of tactical nuclear weapons owing to the numerical disparity in troops between NATO and the USSR and I spent more days of my life than I care to remember, running around the freezing North German Plains, firstly as an Army Reservist, then later manoeuvring around the same Plains as a Regular Infantry Officer in a Mechanized Battlegroup, preparing for a big day which never came. By chance, I was in Berlin on business shortly after The Fall of the Wall when the perceived existential threat posed by the *'Evil Empire'* (President Reagan's 1983 speech) receded at last and hopes of a world peace dividend rose. UK Defence Procurement Minister Alan Clarke wrote at the time:

'We are at one of those critical moments in Defence policy that occur only once every fifty years.'

Feckless optimism, I fear. See where we are today!

It was apparently time to reduce national Defence budgets and the ensuing Defence cuts were deep, controversially so, and in many cases worryingly irreversible. Programmes were shut down, military aircraft and equipment orders were cancelled, and Service personnel numbers were reduced. This all placed intense pressure on Defence Contractors, poor souls, and brought deep joy to Peaceniks everywhere. An essential wave of consolidation survival broke out across the Defence Industry as major Companies hunted for acquisitions to broaden their portfolios and sustain revenue streams.   World defence spending slumped by almost a third.

The remaining British State-owned Arms entities were privatised, as international competition became even more cut-throat. In this febrile atmosphere it was unsurprising that UK Companies particularly, continued to use Arms Cash mechanisms to secure orders, and the late seventies, eighties and early nineties were perhaps the golden age of Arms Cash deals. It certainly seemed that way.

Additional pressure was felt by the UK Defence Industry as a whole raft of States, which had once bought Arms exclusively from Britain for historical reasons, now found their independent feet and climbed into bed with alternative national Arms Suppliers, whilst successive UK Governments were easily out-manoeuvred by more agile competitor Arms-supplying Governments. Simultaneously, the number of international Suppliers of Arms grew as countries such as India, China and Turkey became effective Defence Export competitors Coupled with UK's sudden distaste for manufacturing Industry, it became a perfect storm.

Following the more recent invasion of Ukraine, Western Governments are hastily revisiting their national Defence budgets and the US change of posture vis-a-vis NATO has turned-up the volume of the wake-up call. When British Defence Industrial strategy was modified last time (most of us didn't realise there had been a strategy) overseas Defence Companies were advised to establish their corporate presence in the UK in order to participate in UK Defence opportunities, and the methodology adopted by most of them was simply to buy existing British Defence or Aerospace companies. A foreign buying-fest ensued.

Most long-established UK Defence Brands disappeared forever. French Company Thales, still in c.30% ownership by the French Government, bought Racal, Shorts Missile Systems, Thorn EMI, Helio and Pilkington Optronics, although it was, and remains, impossible for any UK Company to do similarly in France. US Companies also established major UK-onshore presences with Boeing, Lockheed Martin, Raytheon and General Dynamics all setting up shop alongside an increased Italian footprint from Finmeccanica/Leonardo.

The Daily Mail reported that former Defence Secretary Geoff Hoon took up a highly paid job with Augusta Westland, which had landed a £1.7bn UK MoD contract whilst he was still Secretary of State for Defence. Hoon had once been recorded offering to help a lobbying firm by turning the contacts he had made whilst in Government into '...*something that, bluntly, makes money...*' as he put it, suggesting a £3,000-a-day fee for his services. In a completely unrelated and unconnected later development, Augusta-Westland was charged with Arms Cash offences by an Italian Court over various deals in Sweden, Asia and particularly in India.

I've referred throughout this book to 'British-based' Defence Companies and not to 'British Defence Companies', because the majority of top-earning Defence and military aerospace companies in the UK today are in majority foreign ownership, including seven of the top ten UK MoD contractors by contract revenue. The present ownership scenario ensures that massive sums of British Taxpayer's money pass to foreign Defence Company Shareholders and that major job losses and job transfers, plus the loss of technology leadership, have occurred, and continue to occur in Britain.

Overseas Owners dictate what their UK-based Companies can export and often give preference to their domestic subsidiaries for international Sales Prospects. The outflow of IPR, jobs and cash has been massive, and it continues unabated, apparently unnoticed by Politicians. Successive UK Governments boast with monotonous regularity about the value of British Arms Exports to the Economy, a boast which itself is a good indicator of a political moral vacuum.

Now the Government asserts that Defence Manufacturing is an 'Engine for Growth'; how depressing is that? Politicians seem oblivious to the fact that the primary beneficiaries of these 'British' Defence sales are foreign Shareholders and that cash also leaves the country in the form of parent Company 'marketing contributions' from UK subsidiaries to their foreign Parent Companies.

Key IPR and technical capabilities also flee, especially in aerospace and missile sectors, and Governmental naivety on these issues is staggering. No other Nation in history has implemented the self-destruction of their national defence capabilities in such a comprehensive manner, and on such a grand scale, and British politicians seem to be completely unaware of the acute damage they have visited upon the Defence Industry and the wider Economy as a result. I can testify that other Countries laugh at us because of the sheer stupidity of it all.

"Never before in the field of Defence Industrial capability, has so much been thrown away so quickly by so few for so little."

UK Governments like to trumpet the value of Arms Exports and boast about UK's international 'ranking' as if Arms Sales were some sort of fun sporting event, but these boasts are empty echoes from another era, not simply because of the huge cash outflow to foreign Company Owners, but also because of the massive outgoing payments made by the UK to the US Government for Nuclear weapons, and the additional new losses incurred through massive Defence 'gifting'.

These factors combine not only to reduce the value of Arms Export revenues, but totally eclipse them. It would be more realistic to calculate a UK 'Defence Balance of Payments Deficit' figure by factoring in everything that is spent and lost by the unwitting UK Taxpayer on Defence, including criminally mismanaged failed MoD projects. In another classic moral irony, Britain also exports Arms to one of the richest countries in the world, enabling that country to kill people in one of the poorest countries in the world. So much to be proud of there then.

UK also exports Arms and related kit to various countries with lamentable human-rights records and helps Israel to kill Palestinians. Where is *that* league table; the league table of lethality? Like a dog scent-marking its territory, UK Prime Minister 'Stormin' Starmer loves to be seen on TV strutting around in combat kit or riding on a Nuclear submarine, echoing the 'Defence engine for growth' mantra as an apparent route out of national financial trouble. It's almost an exact echo of the '60s Wilson's Labour Government's lunge at Defence exports to get it out of trouble and take note of what happened next. Without those subsequent events, this book would not exist, but now I sense there will be further chapters.

Postcard from the Corporation Nation

Defence Sales, whether for hard-core products like guns, ammunition and missiles, or soft-core items such as Command and Control Systems or even sub-soft support Services, all combine ultimately to facilitate a national lethality capability in the Client country. Sellers have no control whatsoever over the subsequent use of these items, regardless of what any Contract wording might state. Watch that space in Ukraine in the coming years.

Defence Companies insulate themselves from the grief and misery they create with self-delusion and jargon to soften reality. A BAE Systems mission statement included this phrase:

'...We provide innovative solutions and critical capabilities to give our customers an essential edge...'

Anyone know what that means? No references there to the victims of that *'...essential edge...'* A Thales Group Presentation stated:

'We help customers master decisive moments by providing the right information at the right moment.'

This presumably includes the data on civilian targets identified by the Thales supplied Drone Systems in Afghanistan. A Lockheed Martin Mission statement reads...

'We solve complex challenges, advance scientific discovery and deliver innovative solutions that help our customers keep people safe.'

That's a very reassuring slice of Motherhood and Apple pie. But there is no reference to the lethality enabled either by the Corporation's products and systems, or to Lockheed Martin's well-documented and extensive history of Arms Cash activities. As far back as the seventies the Washington Post reported how the Company had:

'...inaugurated and directed a program of foreign bribery that included questionable payments of up to $38 million from 1970 through 1975.'

Historically, Lockheed Martin's Arms Cash locations have included Japan, Italy, Netherlands, Italy, Saudi Arabia etc., and in the USA, Senator Frank Church's Committee concluded way back, that Lockheed Board members had paid $22 million in bribes to foreign officials. It was Lockheed Martin's behaviour that prompted the origination of the US 'Foreign and Corrupt Practises Act' (FCPA) back in 1977.

Lockheed Martin is important because it's now the world's largest Defence Company and has critical mass in countries where its dominance in national Defence inventories makes it indispensable to Governments, regardless of its behaviour. The global annual turnover of Lockheed Martin is greater than the GDP of c.80 Sovereign States and reflects the scenario that the world's major Defence Corporations are more akin to independent 'Corporation States' than they are to being merely Companies. They don't have to conform anymore if they don't want to, and Fines and enforced repayments for malpractice don't make a dent. Lockheed Martin repaid hundreds of millions of dollars to US Authorities for overbilling on contracts, including $265m for overbilling on the F-35 Joint Strike Fighter, as also purchased by UK. The company described this overbilling as 'Inadvertent.' Really? Defence Companies have such a way with words, don't they?

In a recent Stockholder Proposal to issue a 'Human Rights Impact Assessment Report' at the Lockheed Martin Annual Meeting of Shareholders on April 21st, 2022, the Proponent's submission included these words (full text at PX14A6G1j419220px14a6g.htm) and this statement pre-dated the invasion of Gaza by the Israeli Defence Forces:

'Lockheed Martin has provided weaponry such as F-16 fighter jets, Longbow Hellfire missiles, and AH-64 Apache Longbow helicopter parts to the Israeli military, whose occupation of Palestine has displaced over 1 million people and is recognized by the UN as an apartheid. This occupation and apartheid have been connected to 'inhumane acts, arbitrary and extra-judicial killings, torture, the denial of fundamental rights, an abysmal child mortality rate, collective punishment, an abusive military court system, and home demolitions,' and is being investigated for *crimes against humanity. Lockheed Martin weapons have been used repeatedly by the Israeli military on densely populated civilian areas, resulting in thousands of civilian casualties, large groups of them children, potentially amounting to war crimes. Lockheed Martin played a critical role in the May 2021 attacks on Gaza, where apparent war crimes were committed, including the deaths of at least 129 civilians, of whom 66 were children.'*

At this point, I'll re-quote here the Lockheed Martin Mission statement (as at time of writing):

'We solve complex challenges, advance scientific discovery and deliver innovative solutions that help our customers keep people safe.'

Yes, of course you do.

Major Defence Companies echo each other with high-sounding Mission and Vision statements to self-align with all that is right and good. I'm not judgmental about the wording they can say whatever they want, but if you work in the Defence Industry you should really be aware of the reality too, and of the hypocrisy you both reflect and enable. I know from extensive personal experience that few Defence Company employees ever pause to think about this, I certainly didn't, or if they do, they just blank it out. They mostly come to work just to earn money and pay bills like everyone else, but the product of their daily corporate effort enables death, however successfully they or their Employers camouflage or ignore the fact.

In Britain, Defence Companies are legally mandated to train Staff on Bribery and Health and Safety issues but have no obligation to make employees aware of the human cost, the misery of what they enable with their systems or products. The 'Health and Safety' mandate is adhered to slavishly in Corporate Britain, so why not extend that mandate to the victims of Corporate products, or does that element have to remain a well-kept secret?

Chapter 9
Reflections

I tiptoed into the Defence arena when I joined the UK Ministry of Defence as a naïve teenaged Clerk and I enlisted in the Army Reserve Forces in the same month. The Civil Service was my first real (!) job and the MoD was my only Department; I worked in Navy, Land and Air Branches. I was young and glad to have a job, any job. In those far-off days Civil Servants regarded themselves as being a cut above the population at large and MoD Civil Servants certainly regarded themselves as being a cut above all other Civil Servants too.

I passed competitive Civil Service exams, gaining promotion to Executive Grades which carried me to London where I became a Defence Contracts Officer and my Contracts-Branch sidekicks automatically distrusted every Defence Contractor they dealt with. When the MoD imported Sir Derek Rayner of Marks & Spencer's Retail fame to pep-up MoD's Procurement Act, Admirals protested, and Generals gasped. How could the M&S Retail approach to lingerie procurement read across to Guided Weapons?

Little of any consequence changed for us, except that all our contractual draft documents, instead of being completed on site in Southwark, London, were sent instead in little green plastic bags to Brighton for completion and often got lost there. But there were no little green plastic bags to protect Taxpayer's cash, which poured down the drain in irresistible torrents.

The novelty of daily commuting to London and the claustrophobia of bureaucracy soon wore thin though; the work was mind-numbingly dull. I struggled as a Civil Servant because I was seldom civil and never servient, so I didn't really fit in very well. Something had to change. I therefore converted my hobby into my day job and left the MoD on six months Special Leave, to transfer from the Army Reserve Forces into the Regular Army for Operational duty, to became instead, a professional User of Defence equipment instead of a Buyer. My MoD HR Branch advised me that this was a very bad career move and many MoD Civil Servants recognised no connectivity or relevance at all between their day jobs and the Armed Forces.

I volunteered for operational duty in Northern Ireland and served firstly at the School of Infantry and then with a Mechanized Battlegroup in Germany, before deploying to Londonderry. It was supposed to be a six-month Regular Army attachment, but I got carried away and served for nearly four years before returning to the Reserve Forces. I'd never wanted a 'forever' Military career, just as much travel and excitement as I could get. I never returned to the MoD. Militarily that should have been that for me, but it wasn't, as I was seduced by an invitation to try Desert soldiering and I joined the Sultan's Land Forces in Oman.

When I finally ditched my army boots after two years in Oman and a further tour with the Reserves, the Sirens sang again and drew me this time into that 'Hall of Mirrors' which is the world of Defence Sales. I'd bought Defence equipment for the MoD then used it operationally in two Armies, and now I was going to sell some. What was the big attraction?

The first factor was the practical matter of earning a living. I needed a job and knew my way around the Defence Industry where my skillsets gave me a value I couldn't cash-in elsewhere. Well, I used to think not. Secondly, Defence Sales offered excitement in both business and technology senses. There were exciting deals to be chased down on an international stage, notwithstanding the occasional supra-excitement of death threats or bouts of debilitating illness of course, and Defence systems dance on the leading edge of technology; you get to see many tech' things first.

Above all else I couldn't resist the travel opportunities. I would have sold rubber Elephants to get those. The Industry fed my wanderlust intravenously and enabled me to visit places I'd never have reached otherwise. Some have been exotic and fascinating locations to which I would return happily in a flash, but it would take wild horses to drag me under sedation back to others. The balance between the good and the bad has usually been positive though and combining paid foreign travel with an off-beat, beneath-the-radar sort of role was irresistible.

I've witnessed major geo-political events, sometimes too closely, which have buffeted and stimulated the international Defence Industry, and I've needed to adjust to the gigantic technological leaps realised through digitisation which have reshaped Arms and the Defence world, creating unbridgeable technological voids between the USA and the rest. I long ago consigned to the waste-bin the patriotically and oft quoted 'Selling for UK Inc.' mantra, which was so prevalent in the Industry and the Government when I started selling

Defence products, and I remain shell-shocked by the ineptitude and waste wrought upon the UK Armed Forces and UK Taxpayers alike by the MoD. I'm relieved to have been able to focus mainly on Defence export sales.

Truly successful International Defence Salesmen are few and far between and are not the well-balanced psychometrically processed clone team-players so sought-after by today's Corporate HR Departments. Far from it. Instead, they are likely to be strong-willed, arrogant, creative, highly independent, secretive, competitive and probably a bit disturbed. Defence Salesmen reading this description will probably be pleased to be described as such, which pretty well illustrates the point. Sensible, well-adjusted people don't want to do this sort of work and the Defence arena is awash with unengaging and dysfunctional personalities, and yes, I do include myself in this category.

When overseas you should expect to be approached by a veritable army of similar, strange people, all looking for deals or looking for Arms cash; some friendly, some definitely unfriendly, but all with secret agendas. If you're doing the job properly expect to be overseas frequently and often at short notice, and beware, strange things can and will happen in the Hall of Mirrors which your Partner, if you still have one, may not always understand. International travel can be absorbing but it's still just the decorative part of the job and it can be a real killer sometimes.

Corporate monsters at home have to be managed too, since like all other major Industries the Defence Industry is handicapped by high-level corporate politics which often (usually) triumph over purpose, and the Secrecy inherent in the Defence Industry ensures that you'll never really have the full picture. If you're misfortunate enough to be part of a mega aircraft deal or similar, then life can become quite dull quite quickly, as Corporate process and 'teams' take over. Better to find mid-range deals which can be handled alone at the Customer interface.

The leading Defence Corporations have become multi or trans-national entities, with major, often US, Institutional investors. Hence their new pressing need to be 'compliant' or at least to give the impression of so being. Be ready to report to a Boss in a different country with a different business culture and a different interpretation of business 'Ethics'. Some Defence Plc's are disabled by uber risk-aversion, owing to their newly discovered desires to demonstrate 'compliance', whilst others have chosen different paths. Never expect a level playing field. You may find yourself in a foreign country completely unaware that some secret Corporate agenda is playing out in parallel without your knowledge, whilst in other countries, you may be scripting that agenda yourself.

That the Great God 'Compliance' now stands supreme in many Defence Primes, might sound surprising in an Industry which historically took major risks with covert Agents and Arms Cash arrangements, but it's a backlash, a new breed of Lilywhites run the block now and Corporate Process-Junkies have displaced the Buccaneers who used to bring home the bacon.

So, when one of these major Corporations does transgress, it's likely to be on a truly macro scale, involving the highest-level contacts in Governments, Ministries or Ruling families. Remember, not all the world's Defence Suppliers are signed-up to Western business values, so in some situations you will be competing with State-owned or State-directed Businesses with primarily political agendas, so all your wonderful technical data, pricing stuff and compliance undertakings will be of secondary or tertiary importance. Putting that to one side, I'd like to reflect here on some key Defence Industry issues and characteristics, starting with...

Secrets

The Defence Industry is awash with secrets. It creates, nurtures, values and steals them, and maybe secrecy is part of the attraction for people like me, for my own life has been shaped by secrets: Official Secrets, Military Secrets, Trade Secrets, Commercial Secrets, Arms Cash Secrets. The British love secrets, even if they are useless at keeping them, and ever since Lord Walsingham became Elizabeth I's master of State Secrets, the manipulation of secrets has held us Brits spellbound.

In wartime, this obsession with secrecy is understandable. During World War II, Service personnel and civilians alike were imbued with a strong sense of essential secrecy and there was much to be kept secret. There were secret weapons, secret plans, secret Agents on secret missions, and keeping Secrets became an essential fact of national life. A national culture of secrecy was born, embedded and consolidated. *'Careless Talk Costs Lives'* proclaimed one famous Poster *'Loose lips sink ships'* warned another. Britain was fighting for survival and the population was indoctrinated and cowed by a State propaganda, '*Walls have Ears*', '*You never know who's listening*'. Every citizen was conditioned to guard and be guarded with information as never before.

Cold War secrets took over where WWII secrets left off, and the active pursuit of other Nation's Secrets, especially Military secrets, accelerated and became an obsession. East and West vied for technical military advantage as Secrets got bigger and became Nuclear, and new forms of Industrial espionage blossomed as more and more Defence technology was developed in the private sector.

The preservation of national Defence Secrets and the aggressive quest for other Nation's Defence and economic secrets, reached new levels of sophistication and desperation. When Britain spied it was acceptable, when the Soviet Union spied it was the Evil Empire at work, but Nations spied on other friendly States as well as on their enemies. Spying and secrets blossomed into popular culture with an avalanche of books and films adding dark glamour to the essentially grimy world of secrets. We love them, and so does today's Defence Industry.

I handled secrets regularly during my early working life in the Ministry of Defence, and secrets continued to be a central part of my life during Army Service, when deeper Security Vetting enabled me to share even bigger Secrets. I even created new Secrets. So, it was perfectly natural for me and for so many like me, to migrate into and feel comfortable within yet another world of Secrets, this time in the Defence Industry. Like everyone else, I accepted unquestioningly this clandestine culture as the norm.

The Defence Industry not only thrives on secrets, but it is protected from scrutiny and accountability by them too. It is even protected by Government secrecy when it errs from the straight and narrow and that's just about as secret as it gets. Defence Industry Secrets fall broadly into two categories: the 'what' you sell, i.e. product, system, technology or service, and the 'how you sell it', the mechanisms and arrangements necessary to make a sale. It's perfectly reasonable to sustain a level of commercial secrecy relating to IPR or commercial sensitivity as in any Industry, but today's fashionable calls for transparency in Arms deals collide head-on with this culture of secrecy.

International Defence purchases relate to State security and thus are State Secrets, so why would any Customer country or international

Defence Company want to share such sensitive data? What State in its right mind would want to alert potential enemies by advertising its Arms or equipment purchases for the sake of transparency? Countries operating the highest levels of surveillance technology will already know anyway what their potential opponents' armouries contain, and Corporations in most non-Defence sectors also have secrets, so why should the Defence Sector be any different? The key difference is of course lethality, and lethality has been impacted and enhanced by technology in recent years as never before, and technical advantage can only be protected with secrecy.

Some secrets are double-edged weapons, which can also protect and preserve the Useless and the incompetent, and I've lost count of the number of times I've witnessed some feckless Defence Sales Executive hide behind a veil of secrecy or mystery to screen his lack of Sales progress from scrutiny, or to desperately preserve his 'visiting rights' in his favourite territory. For some strange reason, the Far East is a hotbed for this form of denial, and I've managed several Executives in different Companies who sustained visit their programmes for years into Malaysia or Thailand, with absolutely no business outcomes in sight. Each defended their position with an air of contrived mystery and secret contacts.

I once interrogated an Executive who'd virtually taken root in Malaysia, only to determine that after nearly two years of international traveling, lengthy in-country stays and voluminous Visit Reports, he had no valid business prospects whatsoever. His Reports amounted to fiction, secret fiction yes, but still fiction. 'Knowledge is power' runs the saying, which is perhaps more applicable in the Defence Industry than in any other.

The acquisition and retention of market or country knowledge, or at least the illusion of it, confers special powers on Executives. If you manage an international Defence Sales or Business Development team you need to be a competent Interrogator as well as a good Sales Manager, and there aren't many of either of those around. Making commercial progress with market Intelligence is one thing, but making technological progress is another. During my time in the Hall of Mirrors, the technological leaps have been immense, more so perhaps than at any time in the history of the development of Arms, and the impact of this is multi-layered.

Military Technology – The Widening Gulf

Since the industrialisation of the Western world, a military technology gulf has existed and grown between the West and developing nations. In the Victorian era, this disparity was exemplified starkly by the British Army's infamous use of *Maxim Guns* to mow down spear-carrying Matabele tribesmen. Hilaire Belloc caught the moment in verse thus:

> Whatever happens,
> We have got,
> The Maxim Gun,
> And they have not.

Drones and remote weapon systems used to carry-out extra-judicial killings, often against unarmed individuals and civilian targets, are perhaps the twenty-first century's equivalent of the Matabele experience. The victims probably have no concept as to the nature of the weapons which kill them. It's chilling to consider just how quickly we've passed beyond the initial shock of 'extra judicial Drone killings' to a lethargic unchallenging state of acceptance of them. It's Technology's gift to the world. The current wider use of Combat Drones has changed tactical thinking, but can you find Drone Regiments in the UK Army? No of course not.

Nations now kill people simply because they can, and often for political rather than military/battlefield reasons. Small rich States which can afford highly sophisticated digitised weapon systems have become enabled to take military actions which they would previously have been unable to undertake with their limited conventional forces, and this adds to escalation.

The 'decision-to-action' timeframe with hi-tech weapons is much quicker than with traditional military means. Mobilisation actions for conventional military units are slow and are invariably 'telegraphed' to neighbours, whereas Drone strikes for example, or submarine-launched Cruise missiles unleashed from afar, are not only super-quick to initiate, but may be politically unattributable in the immediate aftermath of a Strike. Digitised weapon systems tilt the balance of power between existing adversaries and they also admit new and additional players to the stage.

I've met buyers of Defence equipment in small countries who are hugely excited by the concept of being able to strike whoever and wherever they please, simply because they can, either within their own country or across national boundaries. But the pace of development of digitised weapon systems has not been matched by any equivalent development of the moral or ethical codes for using them, and existing international conventions are inadequate for controlling 21st century technology. There is simply no provision, either for the control or use of Smart Weapons or of the wider sale of them, other than variable and inadequate national export controls. A word of caution though. The second Hague Convention of 1907 prohibited the use of Poison Gas, but that didn't stop Britain from becoming the 'First User' of lethal Gas during World War I. We'll have to improve on that before technological and moral gulfs diverge further, but don't hold your breath.

There's not merely a single Defence technology-gap to think about today though, as there are now several varieties of such gaps to worry about. One is the broad gulf which exists between the West generally and the developing World, whilst a second big-but-narrower gap exists between the West and Russia/China, with a sub-gap opening between those two countries. Perhaps the most concerning for Western Politicians though, is the widening gulf between the USA and her own Allies. This accelerating US weapons-technology lead should concern Politicians everywhere more than many other things, as it will become unbridgeable. For many it already is.

At some basic levels, digitised weapons technology is becoming less controllable and more widely accessible to any developing Nation, or even to groups within Nations. For example, most countries can now operate simple combat or improvised combat Drones (ICDs) with basic sensors or weapon packages. Even a few years ago, no Observers would have expected Houthi irregular Forces to be capable of attacking ships at sea with Drones, but they are now.

Few would have predicted Iran as a major source of Military Drone production, but it is now, and few analysts predicted high-volume Drone warfare exchanges between Ukraine and Russia. Today, Arms Cash changes hands for Digital know-how and not just for bullets and bombs, and the essential digital technical expertise is not available solely from the Defence domain where it previously enjoyed secrecy.

In this respect the military technology playing field just got a whole lot wider and a whole lot less controllable.

The ex-Warsaw Pact States which came in from the cold to join NATO and the EU have been sucked into an expensive vortex of accelerating Military Technology. You want the NATO security blanket? You must buy US high-tech equipment. It's a licence to print cash, Arms Cash, and the geo-politically most vulnerable will pay. Nothing much has changed for Eastern Europe in one respect though, since it's still a Buffer Zone. It had been the Soviet Union's Buffer, but now it's wearing the NATO invisibility cloak as the USA's Buffer Zone. The more States which join NATO, the more Customers there are to pay US and Western Defence Manufacturers for the privilege. Even if the USA ultimately pulls out of NATO, US weapons technology will remain on most States wish lists.

Today's 'must have' hi-tech military systems were first exposed and demonstrated to a stunned world-wide audience a few years ago. CNN's TV coverage of General Schwarzkopf's Gulf War Press Briefings included images of US 'Smart' weapons smashing through designated windows in enemy buildings and exploding inside. That changed everything. Defence Industry Insiders were familiar with the development of these new weapon types, but the wider watching-world was stunned by the precisely delivered deaths now available.

The arcing trajectories of these precision weapons caught not just the public imagination but the attention of National Arms buyers too, especially the rich ones in the Middle East. The televisual treat changed the Arms business overnight, and I hope the US Defence Industry thanked CNN for its supreme marketing efforts on their behalf. It was history's most effective international Arms advertising campaign. US Smart Weapons and their delivery mechanisms became the only game in town.

Small States realised that once armed with high-tech digital weapon systems they could 'participate' in international interventions where previously their conventional Forces would have been inadequate; Defence technology enables more people to kill more people. All the blah-blah about 'Surgical Strikes' is just that, blah-blah; more people die. Countries lacking a full indigenous Defence Technology-base will never approach US levels of digitised major weapon sophistication, as opposed to small Drones etc., so they will always suffer dependencies and vulnerabilities as a result. They will always be paying-Customers, purchasing in small volumes, with regular expensive upgrades to stay current.

A whole batch of potential international conventional Defence product Sales were neutralised by the arrival in the Market of US high-tech weapon systems, leaving non-US Suppliers, who are unable to compete on a technology basis, dependent upon alternative sales 'tools', e.g. Arms Cash. In Defence sales parlance, the phrase 'hi-tech' suddenly equated primarily to 'American'. It was a killer blow for many UK-based Defence Companies, especially those unwilling or unable to team-up with US partners, as I discovered, and my experience highlighted flawed UK attitudes.

I'd been working on a major UK Defence Digitisation Programme with a leading British Defence Company, and I'd approached the world's 'best in class' American Company, which made the in-Service US equivalent with a view to Teaming with them. In the US Army, this equipment was fielded and in-service, but in the UK it remained conceptual.

My simple strategy was to team with this US Company to accelerate the UK Company's position and leapfrog the technical opposition. I flew to the US and made Presentations and found them amenable and willing to team up with us, so I flew home elated, even if the celebratory meal in the 'Oyster Bar' with my new American friends, fell below the high-bar set by my Israeli friend Amnon in Paris with his Lobsters.

With breath-taking arrogance, the Technical Director of the UK Company announced that he could do it better than the Americans, i.e. better than the World leaders, and he set about reinventing the wheel by translating millions of lines of antiquated software code from a first-generation UK system as the basis for his own technical solution. To make this possible, he had firstly to design a tool to translate this old software code before any real Project work could begin.

Inevitably the Company lost the major UK programme it was widely expected to win, and it was gobbled-up by an Undertaker Competitor which stripped-out anything of value and junked the rest. The incident is included here as an example of just how out of touch some UK major Defence Companies had become, a condition enabled by MoD's 'soft contracting' and legacy posture, which ensured that their preferred Companies didn't have to try too hard to win business and many collapsed like powder-puffs when real competitive pressure was applied. Much easier money was available from the UK MoD and the unwitting Taxpayers; why export? Having the Tech' is one thing, but you still have to sell it, and here's another big issue. Not only does the US have the technological lead, it also has the most creative and effective international Arms Sales formulae.

Political Arms Cash

Uncle Sam's 'US Foreign Military Sales' program (FMS) is a formal Government-to-Government Defence policy and support program, which the USA describes as '...*a fundamental tool of US foreign policy...*' so absolutely no confusion about its purpose! Under FMS protocols, designated 'Friendly States' can purchase US Defence materiel directly from the US Government, which in turn buys the equipment from the Manufacturer(s), which provides the Buying Nation with Government-Government reliable, realistic pricing (!) and some assurance as to the absence of Arms Cash deals.

Concluding an FMS deal brings the Client State into an intimate political and military relationship with the USA, and in cases where major systems are sold, aircraft for example, the support, upgrade and revenue relationship between Supplier and Customer is likely to continue for 25-30 years. The Client State will become wholly dependent on (and addicted to) the USA, with all the political implications such a dependency brings. It is akin to the relationship between Drug User and their Dealer. Britain has no equivalent programme.

FMS deals are great propositions for US Defence & Aerospace Manufacturers, who need not undertake the long-running and expensive marketing campaigns which burden their Competitors. Financial risks for the Manufacturer are reduced too under FMS, because since the Company's Paying-Customer is the US Government.

The US Military Forces are numerically huge, so the DoD buys more of everything than most other world Defence Ministries, and DoD procurement is structured to deliver to scale, and is very comfortable with high-volume procurement, which demonstrably is not the case in UK. For example, the US Army operates c. 4,000 Bradley Infantry Fighting

Vehicles (IFVs) whereas UK operates only c. 700 (old) Warrior IFVs. Over 8,000,000 US M16 Rifles and their derivatives have been manufactured, compared with c. 380,000 British SA80 Rifles. 4,800 US Abrahams Tanks were produced, compared with 157 British Challenger 2's in-service and 148 Challenger 3's projected at time of writing. These US huge volume differentials facilitate highly competitive unit pricing for export Customers, especially for high-ticket items like Fighter Jets. There are over 4,600 F.16 jets in service around the world in 25 different countries, whereas there are less than 600 Eurofighters expensively made by a politically engineered European Consortium. USA simply has critical mass in Defence manufacturing. The US 'Defence Security Co-operation Agency' (DSCA) which enables, supports and implements US policy in this area, defines its Mission thus:

'DSCA's Mission is to advance U.S. defence and foreign policy interests by building the capacity of foreign partners in order to encourage and enable allies and partners to respond to shared challenges.'

DSCA is a very substantial US Governmental organisation with c. 20,000 personnel and a budget of c $1.7bn. To put that in perspective the UK Trade and Industry 'Defence and Security Exports' Department website indicates c. 100 employed personnel in London, and a further 27 around the UK, plus a small number of overseas Staff, and not even the Spartans could have been tempted into action by those unfavourable odds.

In addition to FMS, the US 'Foreign Military Financing' Program (FMF) provides loans and grants for foreign purchases by acceptable Foreign States, and the US 'Excess Defence Articles' program allows US Defence materiel to be sold to selected States at reduced prices (i.e. profitable dumping). At time of writing 189 countries are participating in US FMS/FMF and related programmes and it's hard to compete with that. UK Governments and UK Politicians simply do not think or act on that scale.

Outside these various US Government programmes, International Customers can of course make direct purchases from US Defence Companies. UK has no equivalent export mechanisms to FMS or FMF, so a Salesman like I was, representing a UK-based Defence Company, can find himself competing directly with the US Government. I have competed against US Defence Company

Salesmen who travelled to the same Country as me, but bypass the Customer and go straight to their Embassy where they connect with their Military support Staff, who are commercially trained with Career paths in defence export matters.

I wouldn't waste any time trying this approach with the bungling amateurs who sit in British Embassies. I don't make this comment sarcastically, but in my own experience they are very amateur, and they constantly bungle; they also usually lack critical Customer-base contacts and basic business savvy. Other foreign Governments also use Defence exports for political leverage, famously Russia. Vladimir Putin described Russian Federation Arms exports as '*...an effective instrument for advancing our national interests, both political and economic...*' and the establishment of the *Rosonboronexport* organisation was evidence of Russia's pre-Ukraine invasion focus on improving Arms sales. The Russians are not alone in assertively trading Arms for Influence of course, and in 2018, French Armed Forces Minister Florence Parly stated that '*...arms exports are the business model of our sovereignty*'.

The French Government holds a dominant Shareholder stake in Thales, France's major Defence Export Company, which has been caught in numerous major international Arms Cash scandals around the world and which also enjoys strategic high-value Defence Contracts and Company ownership in Britain. Some of the world's biggest Arms Cash scandals have been of French origin.

British-based Defence Companies don't enjoy an equivalent level of Governmental support or leadership, so they are frequently outgunned by superior foreign national Defence Sales strategies and mechanisms, which are linked to, or supported directly by, their own Governments. I was in Abu Dhabi once when Austrian President Kurt Waldheim flew in on an official visit with Austrian Defence Company Officials in tow. I was marketing to GHQ Abu Dhabi at that time, but they advised me to go home now that the Austrian President had arrived.

It was a widely held view within the US Defence Industry that Arms Cash mechanisms alone allowed the Brits to achieve Defence sales overseas, and there is some truth in this assertion, and it's been said to me many times by American friends. Of course, it's easy for US Interests to agitate for tougher international anti-Arms Cash legislation when their own Government offers selected Customers soft Government-to-Government deals. The opportunity to ally closely with the USA is a persuasive political attraction. I'd do the same in their place, always give yourself the best chance. I've commented elsewhere in this book that senior level British Politicians, Diplomats and Civil Servants seem to have completely missed the decline of Britain's image, reputation and capabilities overseas; I really don't know what they've all been doing for years and years. It certainly wasn't promoting Britain. Occasionally a few Jets are sold to Gulf Customers, but this is 'small beer' in the great scheme of things, and in any case, dominant overseas Shareholders are now the primary beneficiaries of such sales.

The toughening of UK legislation restricted UK's ability to sell Defence materiel, especially as UK Government Foreign Policy (is there one?) remained so frustratingly supine. The UK Bribery Act was a typically negative modern British response, i.e. introduce something punitive, rather than develop a more creative and positive strategy, an alternative marketing solution like a UK version of FMS for example.

It's hard for British-based Companies to win major international opportunities if FMS or something similar is on offer to the Customer, but now it's equally hard to find a major Defence or Aerospace Company with majority British ownership anyway. We have been out manipulated and outplayed.

Ironically, the USA, with its much-vaunted 'Foreign Corrupt Practices Act' developed its own special brand of Arms Cash problem. The Special Inspector General for Iraq (SIGIR) Stuart Bowen, stated in 2009 that 'corruption continues to plague Iraq' and reported that US investigators had recovered over $1 billion as a result of their investigative actions involving US personnel. SIGIR's activities are ongoing.

A similar situation arose in Afghanistan with US personnel and British businessmen being arrested, tried, and jailed during the US reconstruction support programme in-country. CBC reported extensively on how the US failure to deal with corruption in Afghanistan was a major contributing factor to the collapse of the Afghan National Army and Police Forces.

Legislation changes in various countries have certainly impacted on Arms Cash events, though maybe not always as the legislators intended. Some Arms Cash arrangements simply submerged deeper whilst others morphed in form to such an extent that they are no longer recognisable as Arms Cash. New styles of contracting, including Contractorisation projects and Offset deals, have enabled Arms Cash architectures to be created which do not show up as old-fashioned cash transfers used to, and almost everyone, except in small deals, has moved on. Traceable cash is yesterday's reward, yesterday's Arms Cash. Unless you Trade from Eastern Europe of course.

Those working on the front line of Defence Sales as I once did, should accept the possibility that Arms Cash exists potentially in any deal in which they are involved whether they are entrusted with the details of it or not. The Arms Cash component may be orchestrated at some lofty secret corporate level, and you may be an unwitting but enabling 'Arms Cash Mule.' In any case, Runners will always approach looking for off-stage deals and be clear from the outset, none of these folks play by the rules.

You may honestly believe that Arms Cash deals are off the table from your Company, but many Runners and Agents will not and probably with good reason; there's a lot of history and baggage in the Arms business. In sizeable deals particularly, you will not necessarily know what is going on behind the screen, either corporately or politically, and this can make you very vulnerable. Remember, you are expendable, unlike a good Agent or Runner.

Agents and Runners

I've no idea when it became the norm for British Defence Companies to pay-off Agents with Arms Cash, but the '60's 'Stokes Report' which led to the founding of the MoD's Defence Export Services Organisation, identified and highlighted the practise of paying Commissions to third parties. The UK Government of the Day embraced Arms Cash activity warmly as an unquestioned necessity when pursuing lucrative Arms Exports, demonstrating as much practically in deals such as the Iranian Tank contract.

Judging by successive UK Government's repeated failures to act against the biggest Corporate Arms Cash spenders, that original warm feeling seems to endure in some quarters of Government to this day. Adjustments to legislation ensured that the term 'Agent' acquired toxic status amongst the new breed of Corporate Executives in UK-based Defence Industry Plcs, but the term had been used freely and normally in marketing circles until then. It became Kryptonite overnight and curious alternative terms sprang up. The most comical of these which I encountered was 'Commercial Facilitator'. For many Companies it was hard to break the Arms Cash habit in one form or another.

Identifying and engaging real influence is a sophisticated activity demanding a high level of contact, often at a national level, and it requires a different approach to that of securing a mere Informer's services. Influencers for major deals are trickier to approach than Informers and are certainly more expensive to secure; they will also be targeted by most competing Companies, sometimes with discreet inter-governmental or political undertakings mixed into the bargain.

Influence is expensive but squeezing basic price or specification data out of the lower levels of the Customer's organisation without the Customer knowing, is a specific sub-set activity which is sometimes quite easy and relatively cheap to achieve. Lower level deals can be achieved using Informers only, whereas major deals for Aircraft, Warships, National C4I Systems etc., will not be won without either key national Influencers at work, and, or Political alignment.

The Political alignment factors may or may not include Arms Cash rewards, but sometimes a pure political commitment may be preferred, e.g. if your Country is attacked, we'll come and bail you out. For Arms Cash paying Companies, multiple Agents are used in some deals, but in any case, the optimum scenario is to ensure that your own Products or Systems are specified and selected over others in the first place, with prearranged budgetary approval in your favour. In fact, just like becoming a 'Preferred Bidder' in the UK MoD process. It may be necessary to participate in a faux competition to give the appearance of 'due process' but your Company is pre-destined to win. If you're not doing this your Competitors certainly are, and the ultimate Arms Cash deal is one created between the Executive/Company and the Agent/Country.

Arms Cash percentages, the going rates, varied in my experience according to several factors, the first of which was the status and level of influence of the Agent himself, or in the very rarest of cases 'herself' as with Fatima or the 'Hairdresser' in Indonesia. The second factor related to what was to be supplied, and the third to the gross value of the contract. A pre-agreed commission lump sum sufficed for a low-level informant where basic information only was provided to ease a deal for basic products, but 'Informants' should never be rewarded on a rolling percentage basis. For an Influencer in a major deal even a tiny percentage can equate to multi-millions in Arms Cash, otherwise anything from 1% to 15% in exceptional cases, may be anticipated depending on the circumstances and nature of the Products/Systems, the Region and the range of Competitors.

Much essential technical and commercial input can be gained legitimately from local Procurement Staff, or maybe from Officers conducting Trials, but 'new-on-the block' Competitors will always pay to swing business by targeting and paying-off these people. In the bygone era of the Shah of Iran's Arms deals with UK, Commission levels were modest in today's percentage terms, but inflation, competition and Agent's expectations have caught up since then. Players know now just how much Defence Companies are prepared to pay to win and how desperate they are too, so expectation is now very high. Agents became increasingly 'picky'.

A Customer community or Country can relatively easily brush-off a corrupt military Officer or an Official, but Agent exposure at national Political or indigenous Royal Family levels can have serious repercussions within the countries concerned and for the Company too. In Defence Companies which are poor at Agent management, and there are many of these, the differences between Informers and Influencers often becomes blurred and confused, with Companies mistakenly treating mere Informants as if they were actually Influencers and rewarding or courting them disproportionately with catastrophic outcomes. I've seen it happen. Bandar was just one example.

This sort of confusion, usually the result of Executive naivety or a desire for backhanders, also leads to over-optimistic forecasting within a Company '...our man is certain the deal will go through by June...' and invariably shock and disappointment ensues when the Bid is lost or deferred, as predictably it will be without the right connections. If only they'd known they were managing an Informer, whilst their Competitor had engaged an Influencer. I've even once wished for a deal to fail simply to smack-down the Main Board smugness (especially at Bonus time) which frequently preceded a deal which had been worked through and prepared in the field by others, who took all the real risks to land it.

Agent approaches to Defence Companies were sometimes purely speculative, with the potential Agent hoping simply to catch a wave of general opportunity and be signed-up, whilst others were planned and calculated specifically to engage a particular Defence Company for a specific deal in return for Arms Cash. This also applies vice-versa with Companies targeting Agents. High ranking Officials, Ministers, Royal Family members or senior Officers who are unable or unwilling to make approaches personally to Companies, may instead send out 'Runners' on their behalf to effect the initial contact and to trigger off-stage discussions.

A key element in some of my Corporate assignments was to follow-up these 'Runner' approaches and the leads they promised within my Region. I had to determine by peeling-the-skins-off-the-onion, which potential Agent or Runner approaches were spurious, and which would lead to genuine Defence contracts or opportunities. It also had to be decided if an Agent was necessary at all, since in many countries Agents were/are either not essential or are illegal, so their very use may inhibit progress or even result in a ban for the Company or jail for Executives. Very senior potential Agents invariably had to be handed over to a central Corporate handler.

The halcyon days of clandestine Arms Cash Agents working for Defence Companies may have passed though they have not ended completely, as evidenced by the occasional conviction, which testifies to the continued use of Arms Cash as a Defence Sales tactic. Usually it's the Minnows who get caught in the net, whilst the Arms Cash Whales swim free and far. Judicial diligence on this subject varies greatly from country to country.

In 2016 it was reported that since 2011, the UK MoD made numerous allegations relating to bribery of which 44 were referred for legal action, with four of these relating to bribing overseas Officials. But this data was only released to Parliament by Minister Harriet Baldwin after Freedom of Information Requests were submitted. It is ironic, well hypocritical really, that UK MoD made such a report when at the same time several overseas Defence Companies with public domain Arms Cash track records, held or participated in major UK MoD contracts. Such is the world of Arms Cash.

Some UK-based Companies use anti-corruption legislation as a positive form of insulation against avaricious Arms Cash Agents, i.e. 'We can't do that... it's illegal...' whilst some others have seemingly never accepted the limitation of the Law, as evidenced by post-Bribery Act convictions/allegations. The UK Bribery Act was trumpeted as the world's most stringent anti-corruption legislation and even the US Foreign Corrupt Practices Act (FCPA) includes scope for some minor facilitation payments. But any such Act is worthless unless supported by the Political Will to apply it, and some Defence Corporations undoubtedly go on their way untroubled except by pesky Journalists.

When prosecution looms, some major Defence Companies have elected to plea-bargain their way out of trouble to escape Bribery convictions, and the fact that the UK MoD remains happy to contract with Defence Companies with massive international Arms Cash track records perhaps says all that needs to be said with regard to UK's attitude to, and acceptance of, Arms Cash.

A popular belief overseas is that many UK International Defence Company Sales and Marketing personnel are active members of the Intelligence Community, and this is a hugely unhelpful belief. In some other countries the situation is different, with Government ownership or dominant shareholdings facilitating the active participation of other 'human assets'. The wretched Matrix Churchill affair blunderers in Iraq did little to curtail or dispel such beliefs concerning UK Companies.

In the Middle East I've twice been directly accused of being a Spy, but I think the accusation owed more to the accuser's perceptions of Britain's historic colonial activities, than it did to any serious belief that I really was in the Intelligence community. They were very uncomfortable moments though. In some major Defence exporting Countries it's a different story, with indigenous Intelligence Services being inextricably and deeply linked to, and embedded within, their national Defence Companies. I've only once knowingly used commercially gained Defence information in a formal Intelligence context, and then only to the benefit of the country concerned, Kuwait.

Defence is a politically sensitive and economically important Industry without a conscience, and common elements between the Defence Industry and Intelligence domains are the constant need for secrecy and the equally constant thirst for information. This confluence of criteria attracts Defence Company Control Freaks, who love to accumulate and hoard data without sharing it, and the Defence Industry is awash with these folks. But you can have both secrecy and mountains of information and still not win a contract, as all the data in the world is useless unless it's convertible into revenue. It's not the possession of data which is critical, it is the interpretation and exploitation of it, and this requires a different mentality and a different skillset. In international Defence markets the maxim for both Company Marketing Executives and Government Intelligence personnel respectively should be one of subtlety, independence and invisibility, not one of inter-operability, inter-changeability or interference. In the mirrored halls it can be suicidal to cross the tracks from being simply Commercial to engaging in Intelligence activity and vice-versa.

Some international Sales Executives acquire an addiction to Arms Cash work and become obsessed with secrecy, the trappings of international travel, and the perceived glamour of the Arms business. Their mantra becomes 'I travel, and I know things, therefore I am'. Their need to win or to deliver any actual business slips from sight, whilst secrecy and the acquisition of privileged information becomes their sole focus, an 'end' rather than a 'means'. On reaching this point they urgently need a change of career and probably assistance from a 'Shrink'. I know several such lost souls and most major Defence Companies have them on the payroll.

In my own experience, the least reliable custodians of useful Defence customer information or data reside within the UK Government's own Defence Export support organisation, once called DESO then DSO and now UK (DSE) which leaked like a rusty sieve when I was active in the Industry. DSE notionally supports (lol) UK-based Defence Companies, but as most of the UK's major Defence Companies are in majority Foreign ownership, UK DSE is actually using UK Taxpayer cash to subsidise foreign-owned Defence Companies to win export Orders.

Defence Industry Executives have ever held conflicting opinions over the role and practical value of DSE, as have different Governments. I negotiated one of my biggest Export deals without DSO knowing anything about it, and at a DSO Seminar I asked an MoD Official if it was appropriate for DSO to support foreign-owned Defence Companies with market information gained at UK Taxpayer's expense. The MoD spokesman wouldn't answer my question in open session. '... *can we take that question offline...* ' he asked.

External organisations and observers strongly opposed to the Arms Trade, are quick to point to DSE as tangible evidence of the Government's participation in international Arms Sales, but that flatters DSE. The widespread belief that UK Governments actually sell Arms is incorrect. They do not. In any case, the Government simply gives them away now at Taxpayer expense, though without Taxpayer consent.

Many regarded the old Soho Square-based DESO simply as an extension of British Aerospace's (now BAE Systems) Marketing Department and feared, often with justification, that anything they told a DESO Desk-officer today would be relayed wittingly or unwittingly to another Company tomorrow. I've certainly experienced that. Frankly, DESO could not be trusted with any important information and were basically useless, unless a formal political 'touch' was required, and that was rare.

In one foreign-owned British-based Defence Company for which I worked, a Director-level Executive was employed solely to collate all inputs from UK DSE and send them abroad to his Corporate HQ weekly for analysis, whereupon these UK Sales Prospects were cherry-picked for action by the indigenous subsidiaries of the overseas Parent Company and its embedded national Intelligence Officers. UK Taxpayer-funded Defence Sales Prospect data is similarly fed back to USA, Italy and all to all other foreign owners of UK Defence and Security Companies. It's a freebie from the UK Taxpayer to the Defence world; no Arms Cash charged.

Institutional Arms Cash

With regard to Arms Cash today, I'm realistic. Politicians may heave self-righteous sighs of faux indignation about Arms Cash, but they stand responsible for the simply gigantic misuse of Taxpayer cash which characterises the profligate and incompetent UK MoD year after year after year. This is 'Institutional Arms Cash' on the grand scale and it soars in value way above any Contract Arms Cash payments.

Britain's Politicians opted to surrender control over UK's indigenous Defence Industrial base in a way in which no other State has ever done, and foreign Defence Company ownership dominates the UK Defence landscape and sucks out the money. Britain has placed itself economically, technically and militarily at a massive disadvantage and the cost in jobs, profits and lost IPR is immeasurable. The resultant logistical vulnerability is going to bite Britain very hard one day and no doubt the troops, not the Civil Servants or Politicians responsible, will pay the heavy price. Our Politicians have made us all vulnerable.

Few of the countries I've visited on Defence business can match the eye-watering levels of financial waste achieved by the UK MoD and none of the many other Governments I've dealt with would tolerate it. Let's be generous here and put this down to incompetence; in any other country I'd be looking for a money trail. To hear our tin-soldier politicians calling for increases in Defence spending whilst the MoD continues to waste billions unchecked is just a sick joke. It's an institutionalised form of Arms Cash.

As a young MoD Contracts Officer, I was compelled to work hard to justify any case for buying equipment without a Competitive Tender, and lengthy handwritten submissions had to be prepared for sign-off at a higher level for any 'Single Tender' action. MoD's own data now shows that only 39% of UK Defence contracts are being placed 'in competition' which means that MoD has self-justified the award of 61% of Defence Contracts without any competition; it's a licence to print for the Contractors.

Successive UK Secretaries of State for Defence and their Procurement Officials have presided over eye-watering levels of financial wastage, with an all too frequent parallel failure to equip the UK Armed Forces effectively. In February 2020, Meg Hillier of the UK Commons Public Accounts Committee (PAC) observed on UK MoD spending as follows:

'For the third successive year, the (MoD) equipment plan is unaffordable. This cannot carry on. The department is locked into short-term thinking, wasting taxpayer's money, and has missed two chances to sort things out.'

Only two missed chances? This has been going on for donkey's years. The MoD could act to resolve these deficiencies but chooses not to, why is that I wonder? According to MoD's own figures, they spent £45.9bn of Taxpayer cash in 2021/22 and this is scheduled to rise (at time of writing) to £51.7bn in 2024, with the Prime Minister declaring an intention to raise spending to 2.5% of GDP and now even 3% is rumoured in some quarters, with NATO calling for 5%. Quite where all this money is going to be found is anyone's guess, but we can see from MoD's track record how it gets wasted.

UK MoD cannot manage the mass of Taxpayer cash already allocated to it, as the Public Accounts Committee (PAC) and National Audit Office (NAO) have pointed out, reporting on MoD Programmes which have been in NAO's own words been '...*eroding our national defence and security for years*' and citing the £1bn over-budget Astute Submarine programme and the £2.75bn over-budget Aircraft Carriers as just two examples amongst very many. The second of these two new over-budget aircraft carriers broke down just outside harbour on her maiden voyage and spent more time in dock than at sea, racking up additional multi-millions in rectification costs, and forcing the Secretary of State to deny that the brand new £3bn *HMS Prince of Wales* was being mothballed.

On land, the horribly flawed £5.5bn AJAX armoured vehicle programme, vital to the Army, was described as 12 years behind schedule. Former head of the Royal Navy Lord West of Spithead commented that '*Ajax programme, no matter how much one dresses it up, has been a complete and utter disaster. It has been a real shambles.*' Another National Audit Office (NAO) Report said the project was '*...flawed from the start...*' with the Ministry of Defence '*...failing to understand the scale and complexity of the work it was undertaking.*' What an understatement. In November 2021 it was reported that this armoured vehicle programme, awarded to General Dynamics, had injured more than 300 military personnel who needed hearing tests because of the damaging noise and vibration caused during the Trials, with only c.14 of the 589 units ordered in 2014 actually being built. The programme, stated as crucial to the Army, was described as '*catastrophic*'. How are such events allowed to continue? Why is no one apparently responsible?

A House of Commons Committee reported that late delivery of other hi-tech equipment had forced troops to rely on near-obsolete equipment. Similarly, ancient Royal Navy Frigates were being used because promised replacements were still awaited. The committee was '*extremely disappointed and frustrated*' by the on-going failure of the MoD. In all, the PAC observed on delays amounting to 21 years and affecting

13 major programmes ranging from Transport aircraft and Radars to tactical communication systems. Armoured vehicle upgrades and torpedoes were also affected.

But despite the PAC declaring itself extremely 'disappointed' and frustrated by this abysmal track record, nothing much changed. When the Head of the British Army, General Sir Patrick Sanders, announced his resignation, he likened the British Army's in-service but ancient armoured vehicles to '...*rotary dial telephones in an iPhone age...*' and I can resonate personally with that. The armoured vehicles I used way-back during my own military service in Germany had entered Army service many years before, in the '60s in fact, and are still being used today. Equally, as Head of the Army, I guess the General was well placed to do something about it?

MoD also spends eye-watering sums of British Taxpayers cash via the US Federal Military Sales Programme and over £1bn was spent on FMS purchases in 2021/22 alone. Key items in this huge spend were Maritime Patrol aircraft and Helicopters. The only military helicopter manufacturer in Britain is of course in foreign ownership. MoD also shows a separate huge spending-line directly with the US Treasury, and the UK Trident submarine programmes are US-dependent.

The price of a single Trident missile 'tube' from the US is currently cited at c.£25m and UK sustains around fifty-eight of them apparently, and a simply astonishing £3.36bn (compare: Pensioner's Winter fuel benefit cuts saved £1.4bn) was apparently paid for just thirty-five F.35 aircraft from Lockheed Martin. UK has surrendered the capability to build any equivalent military aircraft, despite an enviable aerospace heritage and pioneering the vertical take-off technology which marks out the F.35. The only major UK-based military aircraft Manufacturer has a majority foreign shareholder.

In addition to the above quoted payments, MoD also makes on-going contract payments to various UK-based but American-owned Defence and Aerospace Companies including Lockheed Martin, Boeing, Raytheon and General Dynamics, plus UK-based French and Italian-owned Defence Companies, including Thales and Leonardo. Foreign-owned Defence Companies are the major beneficiaries of UK Taxpayer Defence cash, and MoD spent 52% of its massive budget with just 10 Suppliers, with seven of those residing in foreign majority ownership.

It's probable that planned increases in the UK Defence Budget will be gobbled up simply putting existing programmes back on track or replenishing weapons and equipment donated to Ukraine. It's just a form of Institutional Arms Cash with MoD handing over UK Taxpayer's cash to foreign shareholders. Of the many different Defence Ministries with whom I've engaged internationally over the years, only the UK MoD is able to get away with this heady level of ineptitude and wastage. Let's hope it's only that.

Several major UK-based Defence Companies are almost entirely dependent upon MoD Contracts and Taxpayer cash for survival, and Babcock at 55% and QinetiQ at an eye-watering 61% dependency on MoD are good examples of high-dependency major companies. Workers in these Companies are virtually Civil Servants in all but name, ironically so in QinetiQ's case since the Company came into being as a result of a highly controversial floatation of Government-owned Defence Facilities.

QinetiQ Company personnel, who were formerly paid directly as MoD Civil Servants, are now funded as company employees via MoD Contract revenues instead; hard to spot the change there. The National Audit Office stated that greater proceeds should have been secured from the 2006 sale of these QinetiQ Defence assets to Carlyle, which had been nominated as MoD's 'Preferred Bidder' as early as 2003. Why was that one might ask? Sir John Bourn's NAO Report on the QinetiQ privatisation stated:

'The top ten managers at QinetiQ received shares worth £107 million at the time of flotation, from an investment of just over £500,000. These returns were greater than MoD had expected...We think that the returns to management exceeded what was necessary to incentivise them.' Genius.

Selected Civil Servants-cum-QinetiQ Executives did very well thank you from this floatation, and became very rich just for being Civil Servants at the right place at right time, but Taxpayers certainly didn't benefit and the Company continues to be propped up today with public cash via MoD contracts. It's just another institutionalised form of Arms Cash. If I'd set up a deal like that overseas I'd expect to be arrested. I've worked with QinetiQ as a Contractor and it's probably the worst, least Sales-focused, most incompetent Defence Company I've ever encountered. I could hardly get out of the place quickly enough. However, MoD not only privatised many of its own Defence Facilities, it also set about contractorising many Armed Forces roles and functions too.

Contractorisation
I can't put my finger on the precise moment when the MoD lamely followed the US DoD down the Defence path to Contractorisation Hell, but UK Forces now employ civilian Contractors in a multiplicity of roles, from Combat zones to Training bases, from Flight Training to Supply depots and at every point in between. Civilian personnel do jobs previously done by Service personnel but for profit, and the employment of Civilian contractors has eliminated many skillsets and job opportunities entirely from the Armed Forces portfolio.

Contractorisation works well in the USA where there is a positive commercial and social cultural ethos around 'Service' in wider Society, and because the huge comparative size of US Forces means that more US Contractors compete for more contractorised opportunities than is possible in the UK. But the US Military functions in totally different economic and cultural universes to their UK mini equivalents, so what works well in the USA does not automatically work well in the UK. Nevertheless, it seems to be a UK MoD trait to unthinkingly copy whatever the USA does regardless of the impact. As there are fewer contractorisation contracts available in the UK Defence-world compared to the USA, it is easier for a handful of Service Companies to dominate the market and reduce competition, which is exactly what has happened to the detriment of the MoD and the Taxpayer alike. By way of contrast, the benefits to the Service Provider Companies becomes clearer with every MoD payment of taxpayer cash.

From a Company commercial perspective 'Service provision' is much cheaper to undertake than manufacturing as no Inventory or Plant is required, and absolutely everything that's required to deliver a contractorised task, staff, ships, aircraft, vehicles, buildings, facilities etc., can be outsourced. All a Defence Service Provider Company actually needs to win a major Defence Service contract is an office with a few 'phones and computers, plus a line of credit somewhere.

Major Defence Service Companies are more accurately termed 'Service Integrators' as so much Service contract activity can be outsourced. This suits today's limp Executives as there is minimal risk compared to Manufacturing, where Plant, Stock and Materials risks etc., have to be managed. The few commercial risks which do exist in Service Provision can be backed-off onto sub-contractors and sub-service providers, so Defence Service Companies simply sit back and count the cash. It's no wonder there was a headlong rush away from manufacturing.

Contractorisation has robbed the Armed Forces of many skills and career opportunities whilst the profits of Defence Service Providers have soared. It's just another form of Arms Cash with the Taxpayers footing the bill. Overseas Contractorisation deals can also create ideal channels for Arms Cash distribution, allowing cash or assets to be transferred apparently legitimately between parties and two steps removed from host Government departments. Anyone spot an increase in overseas contractorisation projects in recent years?

As if basic contractorisation of Defence tasks wasn't bad enough, UK MoD reinforced failure by adding high value 'Private Finance Initiative' (PFI) Projects to the Procurement mix, which allowed the MoD to burn through even more Taxpayer's cash. In 2020, the National Audit Office (NAO) identified 39 operational MoD PFI Projects valued at £9.24bn, and four of the UK's most expensive PFI contracts of any type were situated in the Defence domain. A NAO Report stated:

'... it (MoD) does not always have the robust data necessary to understand the risks it is asking the private sector to bear.' In at least seven projects, the decision to use PFI has been reversed. 'It may be that in some of these cases, the risks were not fully understood at the outset. In one case, six years elapsed before the MOD aborted a planned PFI procurement for Armoured Vehicle Training Services (AVTS) at a cost to the taxpayer of £15 million'.

In the early years of Defence Contractorisation, projects were simply 'Manpower substitution contracts' in which civilian workers replaced Armed Forces personnel to do their jobs; driving, training, servicing etc., but major escalation set-in when MoD discovered the magical world of 'PFI', and complex long-haul Defence PFI contracts were let for periods of twenty-five plus years, over thirty years in some cases, effectively ending competition forever in some sectors, to the detriment of Taxpayers and Armed Forces Users alike. Once contracted, a PFI Contractor is effectively irreplaceable regardless of performance, and performance can be a major issue as the NAO noted in this example:

The NAO also highlight the risk that contractors may incorrectly report performance which would otherwise lead to payment deductions. Staff of BT inflated the number of calls handled in the Defence Fixed Telecommunications System. The MoD subsequently recovered £1.3 million from BT.'

In PFI projects, Contractors and Consortia combine Manpower Services with the Assets necessary to deliver the nominated contractual tasks. These Assets may include buildings, sometimes taken over from MoD, IT infrastructure, aircraft, vehicles etc. The true costs of PFI contractorisation compared to the alternative costs of in-service provision by the Military, are much more difficult to determine than they are with simple manpower substitution contracting. So Defence Procurers experience considerable difficulty in demonstrating, either to Taxpayers or to the Military, that they are getting any value for money or any real benefits from PFI contracts. Often the reverse appears to be the case.

PFI contracting kills off competition for a generation or more, and this deliberate exclusion of Competition is simply another form of institutionalised Arms Cash, with MoD's 'Preferred Bidders' as the beneficiaries and Taxpayers and Armed Forces as the losers. PFI contracting became MoD's equivalent of an individual's use of a Credit Card. You want something you can't really afford so you pay for it with a Credit Card. In the short term you're happy not to have spent a Capital sum and still get what you want, oblivious to the fact that over time, a much bigger sum is actually paid. A Defence Ministry may think it's making a saving by accessing Assets (aircraft for example) via a PFI service contract, but rest assured it will pay through the nose over the next twenty-five-years. Another NAO Report noted that:

'Affordability, rather than value for money, led the MoD to negotiate a PFI deal for the future strategic tanker aircraft. As a result, it (NAO) is unable to conclude that the ministry achieved value for money in procuring the aircraft.'

Prime Contractors or Consortia in large PFI Contracts usually engage and co-ordinate a whole raft of lower tier sub-contractors to deliver the specified Service. These multiple layers generate corresponding layers of profit, since every commercial entity, every layer involved, has to make a margin. The situation is exacerbated when expensive Assets are also provided as part of the service. Many PFI Assets are funded externally by Financial Service Providers, adding additional debt Interest and management cost-burdens to the MoD contract.

In complex Defence PFI deals the MoD has become the hostage of the Prime Contractors, with Military capabilities being compromised when the Service falls short of the contracted target. Contractor's Commercial teams are invariably smarter than their Defence Ministry equivalents, which can find themselves locked-into bad deals for years and sometimes the issues are so extreme that they become public. In another notable example it took a 2019 National Audit Office Report to highlight the critical shortcomings in MoD's flagship contractorised 'Military Flying Training System' (MFTS), to which the MoD tied itself for twenty-five years, but which failed, according to the NAO, to fully deliver the service, despite the multi-millions of taxpayer cash spent on it. The NAO Report stated that *'As at 31 March 2019, 44 out of the 369 planned MFTS courses had been cancelled due to one or other party failing to fulfil its responsibilities.'*

The impact of such failure on the Forces was that they didn't have enough trained Pilots. The NAO Report further highlighted these key facts about MFTS *'£514 million – the amount received by Ascent (the MFTS Consortium) from the Department (MoD) for the MFTS as at 31 March 2019'* and... *'As at 2015, the Department (MoD) had forecast that MFTS would cost £3.2 billion during the 25-year contract.'* £3.2 billion is an eye-watering amount of Taxpayer cash for a training system which had not functioned fully. In this spectacularly expensive case, the MoD seemed also to have missed the technological trend from Piloted to Autonomous aircraft, and the replacement of many future Aircraft missions by Drones, which has saddled the Taxpayer and the Armed Services alike with a massively expensive manned aircraft Pilot training system for 25 years. In any case, did the MoD honestly believe that

Contractors could or would train RAF Pilots better than the RAF themselves could? Who made that decision?

With Competitors excluded for decades and Prime Contractors enjoying embedded and intimate access to the MoD, PFI contracting has become an institutionalised variant of Arms Cash, but PFI Contracts also pose more fundamental Defence questions. If a Nation can't really afford to train its own Pilots for example, without using the PFI Credit Card, then should it really be doing it at all? Why is the UK MoD buying massively expensive military aircraft for billions of pounds from the USA when it can't even train its own Pilots? Military ambitions and economic realities need to coincide, or they fail, and other Projects need to similarly deliver actual benefits to their Service Customers or be canned.

Company Directors, Investors, Accountants, and Banks all love long-haul Defence PFI contracting (MoD's longest PFI contract was scheduled to run for a blistering 39 years) because they know a license to print money when they see one. Consider MoD's spending forecast of £3.2bn on the flawed MFTS programme for example, and that's just one Project. Senior Company Executives running Defence PFI programmes earn considerably more per capita than their Service equivalents did when they delivered the same Services. How on earth does any Defence Procurer imagine they're getting value for money?

A quick inspection of the published Accounts of one leading Defence Service Provider showed PFI Director-level salaries at over £1m plus bonuses with the Service Contractor still declaring profits, whilst the Senior Military Officer equivalent who would have run the same services and Facilities prior to contractorisation, earned less than 15% of this salary. Similar differentials apply across a raft of costs. There's no need for a Defence Company to pay out illicit Arms Cash for Export Contracts when it can earn billions of pounds at home from Taxpayer Arms Cash through inept MoD PFI contracting. Institutionalised Arms Cash.

Defence Minister Harriett Baldwin's Report that referred to Parliament some 44 allegations relating to Contractor bribery and corruption since 2011, might lead one to think that MoD was 'on the ball' when it came to vetting its own contractors, but a quick look at the history of just one PFI Project tells a different story. The parent

Company of a PFI participant Company paid a $28.4m Fine in the USA for breaching the US Foreign Corrupt Practises Act and was compelled to repay multi-millions of dollars for serious overcharging offences on other US DoD contracts. In 2007, the same Company over-billed the US Government by $265m. The US parent of another participating Company paid a $402m criminal fine in the USA in 2009 on foreign bribery charges, with another part of the same Group being investigated by the UK Serious Fraud Office in 2018 on a different matter. Another participant was investigated several times for overseas foreign bribery allegations, coincidentally including another occasion on which it was working with one of its current UK PFI buddies, and in 2018 the same Company was subject of an HSBC Corporate divestment decision, with similar divestment actions being taken in five Countries on moral grounds. I'm sure the MoD is content with the effectiveness of its Contractor due diligence process which allowed UK Defence contracts to be awarded to these companies. Just don't tell the Taxpayers.

Endgame

Today's Defence arena is much changed from that which I entered full of youthful naivety and patriotic fervour many years ago, and I am much changed too. I was eager to battle with international Competitors for 'UK Inc.' but now those same Competitors own most of the UK Companies I knew so well and suck the cash out of the UK Defence sector. The international consolidation of Defence Companies at home and abroad has blurred the old boundaries and has given birth to giant 'Corporation States' which have moved beyond the control of Governments. The Corporate tail wags the Governmental dog.

These Corporate monoliths need not take commercial risks, why should they when a ductile Defence Ministry will cough up Taxpayer cash virtually on demand? Those Taxpayer-funded Executive bonuses, Defence Company Shareholder Dividends and massive amounts of cash lost in Government wastage, have become just other forms of Arms Cash. In practical terms, it's those major Corporations which now decide what equipment the Armed Forces will get, not the Forces themselves.

Everyone employed in a Defence Company contributes to the overall national lethality capability of their Customers. Whatever their Company produces or supplies, from uniforms to training to rations, to transport aircraft and Fighters, to sidearms, to ammunition, to helmets, to digitised combat radios, to vehicles, to guns, to ammunition, to drones, to smart missiles, etc., every single element contributes to the overall lethality capability of the Client. From the post-Clerk to the CEO, from the Sales teams to the Shareholders, all share responsibility for the deaths which occur as a result of the Industry in which they work. It is inarguable. The very term 'Defence' is now a misnomer, as more and more States no longer feel restricted by their former defensive postures, as they are now enabled by digitised weapon technology to intervene and interfere wherever and whenever they choose, where they could not previously with conventional Forces. Now they kill because they can, and the Defence Industry has made this all possible for them. Governments fail to restrict the growth of such capabilities, preferring Defence Sales revenues to morality and this situation is worsening. The new moral vacuum in which Governments, especially the UK, choose to operate, makes Arms Cash look quite innocent by comparison.

A steady, intoxicating intravenous drip of international travel on Defence business carried me to some remarkable places and to some hellish ones too. I've met exciting, though often damaged and deluded people on the way, together with some of the dullest individuals on the Planet. I've won deals, lost deals, seen memorable and tragic sights, and been fortunate to escape still breathing several times. The world is a much-changed place since I entered the Hall of Mirrors and explored it, and it would be impossible to live that life again today. A moment in time has passed. *Je ne regrette rien*, if only because of the showdown in a Ukrainian Museum between me and the little boy who came along too, finally made me look in the mirror and reflect.

'Only the dead have seen the end of War'
Plato

'Only the end of War will see the end of Arms Cash'
Ralph Houston

www.ingramcontent.com/pod-product-compliance
Lightning Source LLC
Chambersburg PA
CBHW071330210326
41597CB00015B/1406